Numerical and Data-Driven Modelling in Coastal, Hydrological and Hydraulic Engineering

Numerical and Data-Driven Modelling in Coastal, Hydrological and Hydraulic Engineering

Editor

Fangxin Fang

MDPI • Basel • Beijing • Wuhan • Barcelona • Belgrade • Manchester • Tokyo • Cluj • Tianjin

Editor
Fangxin Fang
Imperial College London
UK

Editorial Office
MDPI
St. Alban-Anlage 66
4052 Basel, Switzerland

This is a reprint of articles from the Special Issue published online in the open access journal *Water* (ISSN 2073-4441) (available at: https://www.mdpi.com/journal/water/special_issues/hydraulics_numerical_methods).

For citation purposes, cite each article independently as indicated on the article page online and as indicated below:

LastName, A.A.; LastName, B.B.; LastName, C.C. Article Title. *Journal Name* **Year**, *Volume Number*, Page Range.

ISBN 978-3-0365-0956-3 (Hbk)
ISBN 978-3-0365-0957-0 (PDF)

Contents

About the Editor

Fangxin Fang is a senior research fellow at Imperial College London and executive manager of the data assimilation laboratory at the Data Science Institute, Imperial College London. She leads research on advanced computational tools and data science technologies that can help us to manage a safe, comfortable and healthy environment. She has over 25 years of experience in data assimilation and mathematical modelling technologies. Her main original contributions centre on cutting-edge techniques of predictive modelling (machine learning and data assimilation techniques, reduced-order modelling, adaptive observation), where the work is at the forefront of data-centric modelling. The applications are mainly focused on ocean, atmospheric, multiphase flows, and environmental problems. Fang and her group first applied deep learning techniques to real-time spatiotemporal prediction of nonlinear fluid flows. Applications of advanced data-centric modelling techniques will be renewable energy, weather prediction, pollution forecasting, and hydraulic and coastal engineering.

Editorial

Numerical and Data-Driven Modelling in Coastal, Hydrological and Hydraulic Engineering

Fangxin Fang

Applied Modelling and Computation Group, Department of Earth Science and Engineering, Imperial College London, Prince Consort Road, London SW7 2AZ, UK; f.fang@imperial.ac.uk

1. Introduction

This special issue aims at exploring advanced numerical techniques for real-time prediction and optimal management in coastal and hydraulic engineering. Numerical simulations of fluid dynamics have been indispensable in many applications relevant to physics and engineering.

Multiscale physical modelling: The grand challenge in modelling complex physical phenomena is to predict their characteristics and evolution with adequate accuracy and reliability. This has remained an open scientific problem for decades, unsurprisingly given the extreme disparity in the length- and time-scales involved, spanning several orders of magnitude. For example, climate modelling involves both basin scale and smaller scale features such as boundary currents, mixing, chemical interactions and transport, overflows, and mesoscale eddies. Over traditional structured mesh models, the use of adaptive unstructured meshes provides several profound and widely acknowledged advantages [1]. These include: the ability to conform accurately and efficiently to complex domain geometries (for example, bathymetry, a complex cityscape); the ability to dynamically adapt mesh resolution to improve the accuracy of model results, to capture detailed dynamics, or follow the physical processes.

Data assimilation—incorporating information from experiments and observations to reduce uncertainties in numerical prediction [2,3]. To achieve a predictive capability, numerical modelling often needs to be linked with experiments/monitoring. Data assimilation is a versatile methodology for estimating model uncertainties (parameters, model error, initial and boundary conditions, etc.). A variety of approaches have been used to facilitate data assimilation and include statistical interpolation methods, nudging data assimilation, Kalman filter, Ensemble Kalman filter, and variational (adjoint) methods. Optimal monitoring location is the key in data assimilation. Targeted observations represent a subcategory under adaptive observations where data collection is optimized to improve a particular forecast aspect (e.g., energy, health impact). The targeting approach incorporates dynamical information from numerical model predictions to identify when, where, and what types of observations would provide the greatest improvement to specific model forecasts at a future time [4]. Such targeted observations are important as they will allow the most effective use of available monitoring resources.

Data-driven modelling: Numerical models have benefited from the availability of high-resolution spatio-temporal data due to recent advances in measurement techniques [5], which make the studies of more complex flows possible. However, the computational cost involved in solving complex problems is intensive, which still precludes the development in these areas. Recently, data-driven modelling has gained popularity in rapidly predicting nonlinear fluid flows. Various studies have shown that machine learning methods have potential for capturing non-linear subgrid processes. Reichstein et al. [6] suggest that future models should integrate physical-based modelling and machine learning approaches. Such combination will be the optimal way by which physical modelling results can provide

Citation: Fang, F. Numerical and Data-Driven Modelling in Coastal, Hydrological and Hydraulic Engineering. *Water* **2021**, *13*, 509. https://doi.org/10.3390/w13040509

Received: 5 February 2021
Accepted: 12 February 2021
Published: 16 February 2021

Publisher's Note: MDPI stays neutral with regard to jurisdictional claims in published maps and institutional affiliations.

dynamic understanding from the governing equations, while data-driven modelling results may find some patterns that are not expected from physical modelling.

The Special Issue of Water journal comprises five original papers covering the aspects of the above-mentioned challenges. Overall, the authors focused on different topics, such as, multiscale modelling [7], data assimilation [8], adaptive observation [9], machine learning [10], and application [11].

2. Author Contributions

Juznetsov et al. [7] presented the results with a new 3D unstructured mesh finite volume coastal model, FESOM-C [12]. Over existing structured mesh models, FESOM-C with hybrid meshes (quadrilaterals and triangles) is geometrically flexible and allows the resolution of the mesh to be varied (within reasonable limits) in both the horizontal and vertical, letting the mesh contour complex coastlines. Variable horizontal resolution enables the use of coarser meshes in open sea regions but more refined ones in shallow areas to resolve important small-scale processes such as wetting and drying, sub-mesoscale eddies, or sub-mesoscale dynamics of steep coastal fronts. In this work, the FESOM-C model was applied to an area of the south-eastern North Sea and its performance was evaluated by comparing the results against observations. With use of mixed unstructured meshes, the salinity and temperature gradients, as well as frontal dynamics, can be captured well.

Du et al. [8] explored the capability of ensemble Kalman filter (EnKF) in its applications. EnKF has its potential for efficient use on parallel computers with large-scale geophysical models. EnKF techniques are widely used in operational modelling. The EnKF is based on a Monte Carlo approach, using an ensemble of model representations to build up the necessary statistics [13]. A background error covariance is computed using an ensemble of forecasts, with the current analysis ensemble serving as initial conditions. However, the introduction of perturbations into background variables may break the physical conservation law. One of challenges in EnKF is how to maintain the physical features of variables after inducing the initial perturbations. To address this issue, Du et al. [8] proposed a multivariate balanced initial ensemble generation method based on the multivariate empirical orthogonal function (MEOF) method. The Local Ensemble Transform Kalman Filter in combination with the MEOF based balanced perturbation scheme was applied for improving accuracy of atmospheric general circulation modelling. The study case was the six-hour model forecast from 1 January to 31 March 2004. The prognostic model variables were temperature, surface pressure, wind velocity, and specific humidity. The results obtained from their work suggested that the ensembles integrated from the initial ensemble generated from the MEOF-based perturbations maintain a much more reasonable spread and more reliable horizontal correlation than those from the randomly perturbed initial fields.

Fattorni and Brandini [9] presented their recent work on adaptive observation strategy. Due to expensive field-deployed resources, there is a need to optimally place observations that will maximally improve the accuracy of numerical solutions at forecast times. The optimal observation network could be adapted for a wide range of forecasting goals, and it could be adapted either by allocating existing observations differently or by adding observations from programmable platforms to the existing network. Fattorni and Brandini [9] proposed the observation strategies based on singular value decomposition (SVD). In their work, SVD was used for identifying the areas where maximum error growth occurred, and a correlation analysis was used to limit redundant observations. The case study was Double Gyre, a well-known idealized case to reproduce the seasonal and interannual oscillations of the large-scale circulation in the ocean, useful for the climate system predictability. The results indicated that optimal observation strategies can provide effective and efficient data assimilation for improving predictive accuracy.

He et al. [10] explored the potential of deep learning techniques in emulating the process-based Martinez Boundary Salinity Generator in simulating downstream salinity boundary for the Sacramento–San Joaquin Delta of California, United States. In their work, the multilayer perceptron, long short-term memory network, and convolutional neural network-based models were used. The training datasets were from 1991 to 2002 while validation datasets were from 2003 to 2014. The results obtained in this work showed that deep learning neural networks can provide competitive or superior results compared to the process-based model, especially during extreme (i.e., wet, dry, and critical) years.

Lastly, Gao et al. [11] presented their study of transport time scales (TTS) for a Hyper-Tidal Estuary. The water exchange processes and transport time scales are important factors in governing tracer transport, water quality, and the ecosystem of the basin [14]. The study site chosen in this work was the Severn Estuary, one of the largest estuarine basins in the UK, with typical spring and neap tidal ranges of 13.5 m and 6.5 m, respectively. The aim of this study was to investigate the residence and exposure times of the Severn Estuary in parallel, and to characterize the transport time scales for a hyper-tidal estuary. The results obtained from their work suggested that for all flow and tide conditions, the exposure times are significantly greater than the residence times. That is, there is a high possibility for water and/or pollutants to re-enter the Severn Estuary after leaving it on an ebb tide. The fractions of water and/or pollutants re-entering the estuary for spring and neap tide conditions are very high, with 0.75–0.81 for neap tides, and 0.79–0.88 for spring tides.

In conclusion, this special issue presents recent studies on numerical and data-driven modelling and applications to atmosphere, ocean, coastal, and hydrological engineering.

Funding: This research received no external funding.

Institutional Review Board Statement: Not applicable.

Informed Consent Statement: Not applicable.

Data Availability Statement: Data sharing not applicable.

Conflicts of Interest: The authors declare no conflict of interest.

References

1. Piggott, M.D.; Farrell, P.E.; Wilson, C.R.; Gorman, G.J.; Pain, C.C. Anisotropic mesh adaptivity for multi-scale ocean modelling. *Philos. Trans. R. Soc. A Math. Phys. Eng. Sci.* **2009**, *367*, 4591–4611. [CrossRef] [PubMed]
2. Ghil, M.; Malanotte-Rizzoli, P. Data assimilation in meteorology and oceanography. *Adv. Geophys.* **1991**, *33*, 141–266.
3. Kalnay, E. *Atmospheric Modeling, Data Assimilation and Predictability*; Cambridge University Press: Cambridge, UK, 2003.
4. Daescu, D.N.; Carmichael, G.R. An adjoint sensitivity method for the adaptive location of the observations in air quality modeling. *J. Atmos. Sci.* **2003**, *60*, 434–450. [CrossRef]
5. Taira, K.; Hemati, M.S.; Brunton, S.L.; Sun, Y.; Duraisamy, K.; Bagheri, S.; Dawson, S.T.; Yeh, C.A. Modal analysis of fluid flows: Applications and outlook. *AIAA J.* **2020**, *58*, 998–1022. [CrossRef]
6. Reichstein, M.; Camps-Valls, G.; Stevens, B.; Jung, M.; Denzler, J.; Carvalhais, N. Deep learning and process understanding for data-driven Earth system science. *Nature* **2019**, *566*, 195–204. [CrossRef] [PubMed]
7. Kuznetsov, I.; Androsov, A.; Fofonova, V.; Danilov, S.; Rakowsky, N.; Harig, S.; Wiltshire, K.H. Evaluation and Application of Newly Designed Finite Volume Coastal Model FESOM-C, Effect of Variable Resolution in the Southeastern North Sea. *Water* **2020**, *12*, 1412. [CrossRef]
8. Du, J.; Zheng, F.; Zhang, H.; Zhu, J. A Multivariate balanced initial ensemble generation approach for an atmospheric general circulation model. *Water* **2021**, *13*, 122. [CrossRef]
9. Fattorini, M.; Brandini, C. Observation strategies based on singular value decomposition for ocean analysis and forecast. *Water* **2020**, *12*, 3445. [CrossRef]
10. He, M.; Zhong, L.; Sandhu, P.; Zhou, Y. Emulation of a process-based salinity generator for the sacramento–san joaquin delta of california via deep learning. *Water* **2020**, *12*, 2088. [CrossRef]
11. Gao, G.; Xia, J.; Falconer, R.A.; Wang, Y. Modelling study of transport time scales for a hyper-tidal estuary. *Water* **2020**, *12*, 2434. [CrossRef]
12. Fofonova, V.; Androsov, A.; Sander, L.; Kuznetsov, I.; Amorim, F.; Hass, H.C.; Wiltshire, K.H. Non-linear aspects of the tidal dynamics in the Sylt-Rømø Bight, south-eastern North Sea. *Ocean Sci.* **2019**, *15*, 1761–1782. [CrossRef]
13. Evensen, G. The ensemble Kalman filter: Theoretical formulation and practical implementation. *Ocean Dyn.* **2003**, *53*, 343–367. [CrossRef]
14. Uncles, R.J.; Stephensa, J.A.; Smithb, R.E. The dependence of estuarine turbidity on tidal intrusion length, tidal range and residence time. *Cont. Shelf Res.* **2002**, *22*, 1835–1856. [CrossRef]

Article

Evaluation and Application of Newly Designed Finite Volume Coastal Model FESOM-C, Effect of Variable Resolution in the Southeastern North Sea

Ivan Kuznetsov [1,*], Alexey Androsov [1,2], Vera Fofonova [1], Sergey Danilov [1,3,4], Natalja Rakowsky [1], Sven Harig [1] and Karen Helen Wiltshire [1]

[1] Helmholtz Centre for Polar and Marine Research, Alfred Wegener Institute, Klußmannstr. 3d, 27570 Bremerhaven, Germany; Alexey.Androsov@awi.de (A.A.); Vera.Fofonova@awi.de (V.F.); Sergey.Danilov@awi.de (S.D.); Natalja.Rakowsky@awi.de (N.R.); Sven.Harig@awi.de (S.H.); Karen.Wiltshire@awi.de (K.H.W.)

[2] Shirshov Institute of Oceanology RAS, 117997 Moscow, Russia

[3] A. M. Obukhov Institute of Atmospheric Physics RAS, 119017 Moscow, Russia

[4] Department of Mathematics and Logistics, Jacobs University, Campus Ring 1, 28759 Bremen, Germany

[*] Correspondence: ivan.kuznetsov@awi.de

Received: 15 April 2020; Accepted: 11 May 2020; Published: 15 May 2020

Abstract: A newly developed coastal model, FESOM-C, based on three-dimensional unstructured meshes and finite volume, is applied to simulate the dynamics of the southeastern North Sea. Variable horizontal resolution enables coarse meshes in the open sea with refined meshes in shallow areas including the Wadden Sea and estuaries to resolve important small-scale processes such as wetting and drying, sub-mesoscale eddies, and the dynamics of steep coastal fronts. Model results for a simulation of the period from January 2010 to December 2014 agree reasonably well with data from numerous regional autonomous observation stations with high temporal and spatial resolutions, as well as with data from FerryBoxes and glider expeditions. Analyzing numerical solution convergence on meshes of different horizontal resolutions allows us to identify areas where high mesh resolution (wetting and drying zones and shallow areas) and low mesh resolution (open boundary, open sea, and deep regions) are optimal for numerical simulations.

Keywords: numerical modelling; unstructured meshes; finite volume; North Sea

---

1. Introduction

Numerical ocean models are one of the major instruments used to understand ocean dynamics. Their area of application is usually divided into global or open ocean models (with resolutions from a few up to several tens of kilometers), regional models (that include coastal seas and whose scales are typically one to two nautical miles), and models capable of representing estuaries or certain specific processes (with horizontal scales in meters). Different basic assumptions about the physical processes to be included or excluded in a particular case are one of the reasons for these divisions; it reduces unnecessarily complicated equation-solving. For example, tides are commonly excluded from the global ocean models used in climate simulations (as in, e.g., [1]) but are dominant in the dynamics of coastal regions. Another significant reason is a limitation on horizontal discretization: models for larger domains use coarser horizontal resolutions to speed up numerical calculations, parameterizing or disregarding small-scale processes. Such limitations usually result from the finite difference method used to discretize dynamical equations in the most well-known, established models: NEMO [2], ROMS [3], MOM [4], GETM [5], and many others. While the finite difference method yields realizations quickly and easily, it is only applicable to structured meshes, making it nearly impossible to construct meshes with variable resolution that can resolve specific areas of interest as needed (coastlines, archipelagos, and shelf breaks) but which also coarsen towards open ocean (as in, e.g., [6]). Moreover, because of computational demands, the finite difference method would not be employed to resolve big domains (ones the size of a regional sea) at the resolution needed for a coastal process (ten to hundreds of meters) unless some nesting were applied.

Studies in recent decades clearly show the necessity of combining different scales in a single model to address phenomena such as the transport of matter between the coast and the open ocean, the effect of regional processes on global ocean dynamic [7,8], or more technical questions related to regional models, such as open boundary conditions [9,10].

Nesting of two or more structured grids of different resolutions is one of the common approaches used in structured-mesh codes. A widely used one-way nesting method (meaning that information from a lower-resolution grid is transmitted to a finer grid), as in [11,12], shows good results, but the same method cannot be properly applied to explore flows in the transition zone. Two-way nesting (meaning that both grids continuously exchanging information), as in [13–15], is more complicated and incurs difficulties with smoothing and damping of signals between coarser and finer domains. This lets the most important small-scale process be filtered out or left unresolved.

The alternative to structured-mesh methods are ones designed for unstructured meshes. They are geometrically flexible and allow the resolution of the mesh to be varied (within reasonable limits), letting the mesh contour complex coastlines. Such methods are used successfully by a number of well-developed ocean models such as FVCOM [6,16,17], SCHISM [18,19], and FESOM2 [1]. Most of these models are oriented toward regional or process studies. To date, most of the experience with global large-scale application has been accumulated working with FESOM. The importance of regional and coastal model studies on various time scales is hard to overstate, be it for process studies [20,21] or climate research ([22,23]). However, in most cases, there is no dynamical link to global models [24].

With either approach, nested structured meshes or unstructured meshes, the central question is what the optimal horizontal resolution is for a specific task in the modeled region. In other words, where does the best compromise lie between the quality of the simulated dynamics and computational efficiency?

Here, we present the results of simulations performed with the newly developed FESOM-C model [25,26]. FESOM-C is a coastal branch of The Finite-volumE Sea ice–Ocean Model (FESOM2) [1]. The FESOM-C model employs hybrid unstructured meshes [27] and is based on a finite-volume discretization. It is a full three-dimensional model based on three-dimensional primitive equations for momentum, continuity, and density constituents [25]. It includes modules for the open boundary, upper boundary (interaction with the atmosphere), rivers, output, and postprocessing, all of which facilitate using this model in realistic applications. In practice, the hybrid meshes used by FESOM-C consist mostly of quadrilateral elements and include triangles only where needed to link quadrilateral cells. Compared to purely triangular meshes, this significantly increases model throughput (see, e.g., [27]). Variable horizontal resolution enables the use of coarser meshes in open sea regions but more refined ones in shallow areas to resolve important small-scale processes such as wetting and drying, sub-mesoscale eddies, or sub-mesoscale dynamics of steep coastal fronts. The general structure of the model's internal arrays, variable names, and mesh utilities is similar to that of FESOM2. Its modules for external forcing, output, and parallelization are likewise similar. However, several principal aspects render the FESOM-C model capable of representing many of the physical processes of the coastal areas, such as tides and the wetting-drying mechanism. FESOM-C also differs in that it uses a terrain-following vertical coordinate. At the same time, the closeness of both models makes establishing dynamical links between coastal and global realizations relatively easy. A detailed description of the FESOM-C model is presented in [25]. One of the goals of this study was to analyze the capabilities of the FESOM-C model as a coastal model in a realistic setup. We chose to apply the model to an area of the southeastern North Sea (the "SeNS") (see Figure 1), which is a comparatively well-studied area. Moreover, it offers the comprehensive datasets which are essential for model setup and validation.

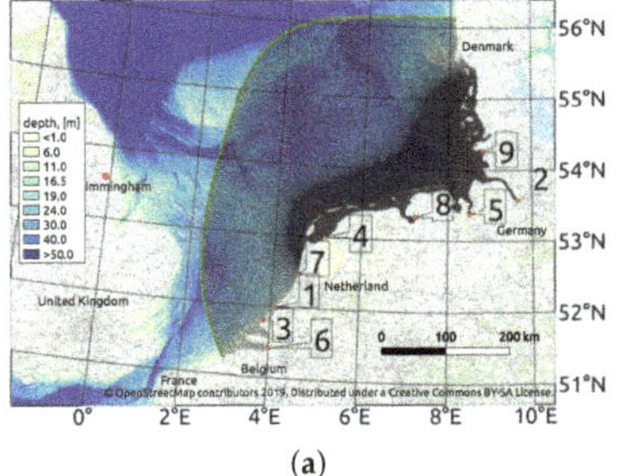 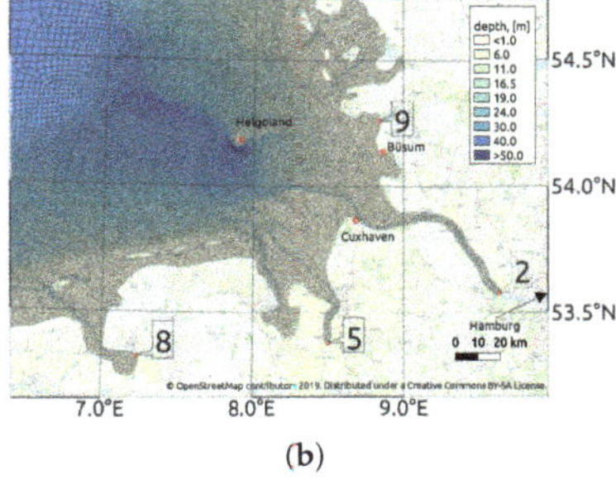

(**a**)   (**b**)

Figure 1. Bathymetry of the southern North Sea (colored contours, data from EMODnet Bathymetry portal); black lines show the mesh used for the five-year run: (**a**) a full domain; and (**b**) a zoom-in to the Cuxhaven–Helgoland area (the southeastern part of the full mesh). The background map was created with openstreetmap.org (© OpenStreetMap contributors 2019. Distributed under a Creative Commons BY-SA License).

The North Sea coast, and the SeNS region in particular, is a densely populated hub with many seaports, including Europe's biggest. Millions of livelihoods depend on the state of the North Sea. Recently-built wind farms cover a significant part of the coastal area [28,29]. The mean depth of the North Sea is about 80 m and the maximum more than 700 m, but depths in the SeNS range from 20 to 80 m. Sea-bed transformation is frequent here. The North Sea is connected to the Atlantic Ocean by the English Channel to the south and the Norwegian Sea to the north. The mean wind-driven circulation pattern is counterclockwise. Tidal dynamics are defined mainly by the semidiurnal principal lunar tide (M2). The M2 tidal wave propagates from the north and the southwest and enters the SeNS region on its western boundary. Of the three amphidromic points in the North Sea, two of them are in the SeNS. Superposition of the M2 and S2 tides causes significant spring-neap tides. The North Sea's physics are briefly described in [30]. There are several significant freshwater sources including the rivers Rhine and Elbe in the SeNS, forming strong horizontal salinity gradients. The Wadden Sea, a series of islands separated by tidal inlets and characterized by significant tidal flats and wetlands, plays an important role in the dynamics and ecological state of the SeNS [31].

As mentioned above, the North Sea, and the SeNS in particular, is a quite well-studied area. Many models have been applied to simulate its dynamics. Recent work in pre-operational modeling has focused on the southern North Sea with a combination of numerical and observational methods, as demonstrated by Stanev et al. [32]. Comparing several models, the authors found a significant difference in a nonlinear effect within the tidal dynamics of the shallow coastal zone, which they attributed in all likelihood to coarse horizontal resolutions. An evaluation [33] of two operational models with relatively high resolutions (up to 900 m) in the North Sea region, BSHcmod v4 and FOAM AMM7 NEMO, found good agreement between the models and observations of the open sea. Near the coast however, the study found a significant deviation between simulated and observed salinity. Haller et al. [33] indicated limited spatial resolution and complicated modeling in the transitional zone as weaknesses of both models. Several one-way nesting system variations, including the TRIM-NP [12] and GETM [11,34] models, have been applied successfully to the North Sea with a focus on SeNS. Gräwe et al. [11] concluded that successful modeling of tidal flow at various Wadden Sea inlets requires various minimum resolutions for some inlets. 500 m may be enough for some, but even a horizontal resolution of 200 m is insufficient for others. SCHIM, which uses finite element and finite volume discretization methods on unstructured meshes, has been successfully applied for the coupled North

Sea–Baltic Sea system with a local resolution of approximately 200 m in the SeNS [35,36]. The same model successfully applied to the Ems Estuary (Pein et al. [37]).

The main objective of this article is to demonstrate the representativeness of the latest results of the first fully realistic, three-dimensional, multi-year baroclinic hindcast simulations with the newly developed FESOM-C model. This is done by comparing model results with various observational data available for the period 2010–2015 and with those of other available models for the SeNS. Another, equally important objective is to present an application of convergence analysis of solutions for grids of different spatial resolution.

The paper is organized as follows. Section 2 describes the model setup for the SeNS. The model results for barotropic and baroclinic formulations are described in Section 3. Convergence numerical solutions for meshes with different spatial resolutions are discussed in Section 4. Section 5 offers conclusions.

2. Model Setup: Southeastern North Sea

2.1. Bathymetry and Mesh

Data from the EMODnet Bathymetry portal [38], with a resolution of about 230 m, were used to construct the model bathymetry. Scattered bathymetry data from the Wasserstraßen- und Schifffahrtsverwaltung des Bundes of the German Federal Ministry of Transport and Digital Infrastructure, with resolutions of up to 1 m in riverine areas, are also employed in this region.

A low-resolution mesh (spatial resolution between 1 and 4 km with 43,318 vertices) and 21 vertical sigma layers were used to simulate the period from 2010 to 2014. This mesh was constructed by the Gmsh mesh generator [39] with the Blossom-Quad method [40] and consists mainly of quadrilaterals. Androsov et al. [25] showed that the quality of quadrilateral meshes constructed by Gmsh, even in the presence of acute angles and degenerate quadrangles, is good enough for solution convergence and the stability of FESOM-C.

2.2. Boundary Conditions and Atmospheric Forcing

For our final simulations, we used data from hydrography reconstructions based on optimal interpolation by Nú nez-Riboni and Akimova [41]. Monthly resolved data were linearly interpolated by the model on the current time step. A relaxation time parameter of 15 days (half the time of the available data resolution) in the case of propagation into the domain, and of 5 days in the case of outward propagation, were applied for temperature and salinity at the open boundary.

Sea-surface elevation at the open boundary was prescribed by amplitudes and phase for the nine (M2, S2, N2, K2, K1, O1, P1, Q1, and M4) most significant tidal harmonics in this area. Data from regional tidal solutions for the European Shelf 1/30° from the TPXO model ([42]) was interpolated onto open boundary locations. High grid resolution (2 km) of TPXO regional setup and improved variational data assimilation method for the shallow-water tides together with carefully selected set of tide gauges gives a good enough solution for area of open boundary. The sensitivity study of M2 wave propagation employed only M2 tidal harmonic data. These data were extracted from Danilov and Androsov [27], who modeled the full North Sea using the previous version of the FESOM-C model. Details of the open boundary implementation are provided in [25,43].

Data from the European Union's Seventh Framework Program (EU FP7) project "Uncertainties in Ensembles of Regional Re-Analyses" (UERRA) [44] constitute the surface atmospheric forcing. The atmospheric data are derived from a data assimilation method that assures its quality. The ocean model utilizes fluxes of freshwater (rain and snow), short- and long-wave radiation, surface wind, humidity, air temperature near the sea surface, and air pressure at sea level. The time resolution of the atmospheric data is 1 h; their horizontal resolution is about 11 km.

2.3. Initial Conditions and Spin-Up Period

Preliminary sensitivity studies have shown that perturbation in the initial temperature and salinity fields is compensated for after one year of simulations in such a way that the model solutions with varying initial conditions are very close to one another. Initial conditions for the one-year spin-up runs were constructed from TRIM-NP model results [12].

2.4. Rivers

Strong cross-shore salinity gradients in shelf areas like the SeNS are mainly determined by the supply of fresh water from rivers. Temperature and observed daily river-runoff, put together by Kerimoglu et al. [45] from Radach and Pätsch [46], were used to prescribe freshwater supply. A salinity of 0.1 [psu] was used for river water to avoid numerical instabilities due to frontal dynamics near fresh water sources. In total, nine freshwater sources were prescribed (see Table 1).

Table 1. Freshwater sources in the current setup based on daily observed data.

River Name	Discharge [m^3/s] mean/min./max./std.
1 Nieuwe Waterweg	1408/11/4044/606
2 River Elbe	783/261/4070/505
3 Haringvliet	546/1/5903/817
4 Lake IJssel	521/1/2935/402
5 River Weser	269/87/1320/195
6 River Scheldt	127/35/615/91
7 North Sea Canal	84/1/365/41
8 River Ems	70/20/372/52
9 River Eider	23/12/40/6

During the simulation period, several "flood" events comprising a significant increase in water discharge were observed during the winters of 2011, 2012, and 2013 and the summer of 2013. Details on the effects of the 2013 summer flood event are described in [47].

3. Simulation Results

In this section, we provide a basic validation of the simulation carried out on a mesh with variable resolutions of about ~4 km in the SeNS area for the five years from 2010 to 2014. As previously mentioned, the SeNS is the most heavily observed area, with the amount of data increasing significantly over the last few years as new instrumentation has been developed. Several databases, such as EMODnet and COSYNA, contain a significant part of this data. Nevertheless, a consistent database and a method for model validation in this area are lacking.

3.1. Tidal Dynamics

Tides are one of the main driving forces in this area. Accurately representing them is one of the most critical factors in any successful description of the coastal dynamics.

Previous work has demonstrated that the model can accurately reproduce tidal dynamics ([25–27]). To test how the current model setup performs at reproducing the main tidal harmonics in long-term simulations, we analyzed observed sea-level height at several regional stations by extracting the amplitudes and phase of nine harmonics for comparison with model results. We also performed a sensitivity run with only the M2 harmonic prescribed at the open boundary.

3.2. M2 Tide

To test the ability of the model to reproduce the main tidal wave (M2 harmonic) in the SeNS domain, we constructed an additional 2D barotropic setup (by switching off the baroclinic part) with only the elevation from the M2 tide wave prescribed at the open boundary. To get a clear tidal wave and make the analysis simpler and more accurate, we disregarded atmospheric forcing for this 2D setup. Analyzed upon onset of an equilibrium regime after several days of simulation, the results of this simulation were compared with observations, as shown in Figure 2. The M2 tidal constituent propagates along the coast of the North Sea as a Kelvin-type wave. It enters the model domain at the western boundary and propagates eastward to the Elbe River estuary and then northward along the coast. The SeNS area is characterized by two amphidromic points (points of zero amplitude): one in the southwest around 3.5° E, 52.5° N, and the other in the north around 5.5° E, 55.2° N. Both of the M2 wave's amphidromic points are well represented in the model (Figure 2). The amplitudes and phases were compared to observed values provided by Ole Baltazar Andersen (personal communication, 2008). The amplitude and phase accuracy is characterized by the total vector error ([25]):

$$\mu = \frac{1}{N} \sum_{n=1}^{N} ((A_* cos(\phi_*) - A cos(\phi))^2 + (A_* sin(\phi) - A sin(\phi))^2)_n^{1/2}, \tag{1}$$

where A_*, ϕ_* and A, ϕ are the observed and simulated amplitudes and phases, respectively, at N stations. For the current model setup and 53 observational stations, the total vector error is 0.21 m, compared with a maximum of wave height of 2 m. This can be regarded as a good result. The most significant error is simulated near the coast. Most of the discrepancy can be explained by uncertainty in the model's bathymetry and bottom-drag parameterization. In general, the model reproduces the observed values reasonably well but with some exceptions. Stations in the area of Cuxhaven (to the west of the domain) show a smaller amplitude than the observational dataset provided by Ole Baltazar Andersen. However, in the experiment with nine prescribed tidal harmonics at the open boundary, the stations in this area (Helgoland and Cuxhaven) are reproduced well by the model, even the M2 amplitudes and phases (Figure 2).

3.3. Main Tidal Harmonics on Long Time Scales

The effect of neap-spring tide variability is known to be important for coastal dynamics. A number of previous studies, both observational and model-based, have shown the importance of tidal harmonics besides M2; in the studied area in particular, tidal non-linearity also plays a role in the dynamics of the coastal areas ([32,48,49]). We prescribed nine tidal harmonics at open boundaries for our 3D baroclinic run for the years 2010–2014. We compared the amplitudes and phases of the nine main ones simulated with FESOM-C, and as observed at four stations, with results from the tidal solutions of the TPXO model, European Shelf 1/30° version. The stations' positions are indicated by the black dots in Figure 2a,b. The observed amplitude and phase values are based on the years 2010–2014 and calculated using the uTide python module ([50]). The stations used to validate the model were selected for their location at points of interest in the study of tidal dynamics and for providing observed values that cover the simulated period. Two stations, K13a3 and Hoek van Holland, lie near one of the amphidromic points of the M2 tidal wave. The Helgoland and Cuxhaven stations are located near the position where the M2-only simulations yielded the maximum deviation from observations. The Hoek van Holland and Cuxhaven stations are both land stations. The amplitudes and phases simulated by FESOM-C correspond well to the observations; comparing the M4 tidal constituents, we find them just as good as or better than the TPXO model. The Hoek van Holland station is exceptional: here, FESOM-C shows higher amplitudes for M2 and M4. Its vicinity is poorly resolved in the current version of the model setup, with mesh vertices approximately 4 km apart, so bathymetric errors play the most significant role here. Moreover, data from this area are assimilated into the TPXO model. However, the results at the Cuxhaven station, where FESOM-C has significantly higher resolution than TPXO, show the opposite: FESOM-C reproduces all the tidal harmonics very well. Direct comparison of FESOM-C results with observations shows high correlations between modeled and observed values. The standard deviations (STD) are quite close for all stations (see Table 2).

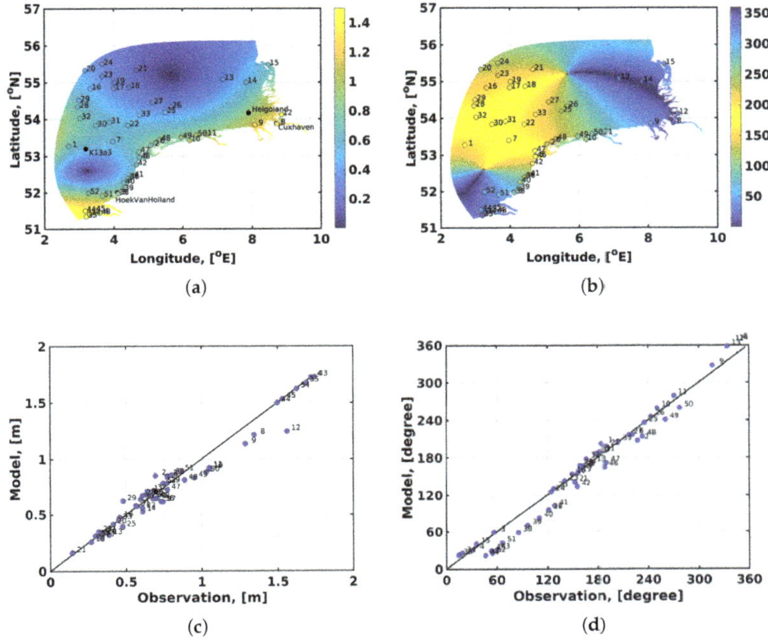

(a)   (b)

(c)   (d)

Figure 2. Comparison of the modeled and observed characteristics of the M2 tidal wave for: the amplitude, in meters (**a**); and the phase, in degrees (**b**). The color maps present model results; colored circles correspond to observations. (**c**,**d**) Scatter plots for amplitude and phase, respectively, for the entire domain. The numbers in the panels are the ID numbers of the stations. The total vector error is 0.21 m. The black circles and text in (**a**,**b**) indicate the positions and names of the stations in Figure 3.

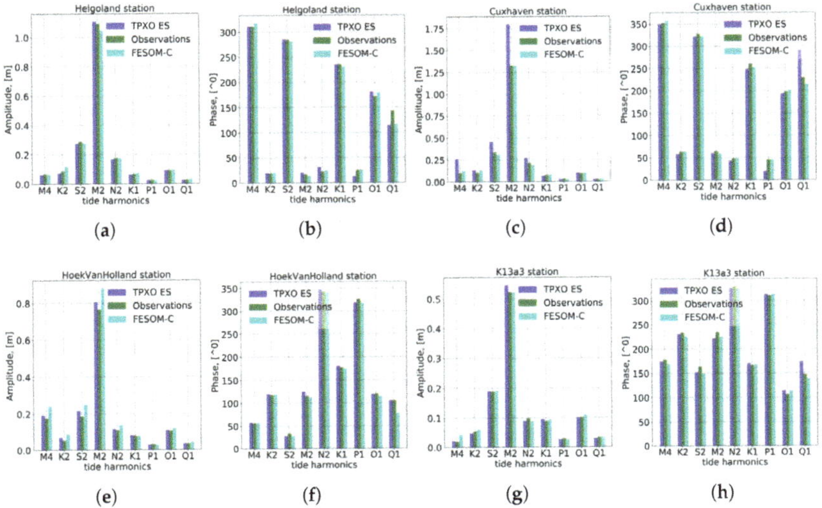

(a)   (b)   (c)   (d)

(e)   (f)   (g)   (h)

Figure 3. Amplitudes (**a**,**c**,**e**,**g**) (meters) and phases (**b**,**d**,**f**,**h**) (degrees) of the nine main tidal harmonics at four stations in the modeled area. Green shows the calculated ([50]) values based on observed sea surface elevation during the years 2010–2014. Blue shows the regional (European Shelf 1/30°) tidal solutions of the TPXO tidal model. Cyan shows FESOM-C model results processed in a way similar to the observations.

Table 2. Correlation between observed and simulated sea surface height (ssh) at four stations. Standard deviation (STD) in time arrays of observed and modeled values of ssh.

Station Name	Correlation	STD Observation	STD FESOM-C
Helgoland	0.95	0.90	0.81
Cuxhaven	0.93	1.11	1.02
Hoek van Holland	0.91	0.67	0.7
K13a3	0.89	0.47	0.42

3.4. Surface Salinity and Temperature over 2010–2014

Modeled salinity and temperature are verified on the basis of data from ICES (Figure 4), COSYNA (maintained by the Helmholtz Zentrum Geesthacht and accessible at www.cosyna.de ([51]), and EMODnet (Figures 5–7) databases. The model captures the observed ([52]) lateral salinity gradient (Figure 4a) reasonably well. Due to residual barotropic currents, a strong horizontal salinity gradient forms along the coast. Figure 4b,c, respectively, provide a comparison of modeled surface salinity and temperature with observed data from the ICES database for the years 2010–2014. With some exceptions, the observations are from locations outside areas with a strong horizontal salinity gradient, in the open area of the simulated domain, where the vertical structure of the water masses is more susceptible to variability associated with a seasonal thermocline. The corresponding Pearson correlation coefficient (Cor.) and root mean square difference (RMSd) are indicated at the top of the plots. Simulated surface salinity and temperature correlate well with observations. The high-temperature correlation of 0.99 can be explained by the seasonal cycle. The relatively small RMSd for both temperature and salinity attests to the model's ability to reproduce the overall dynamics and thermohaline structure of the simulated region.

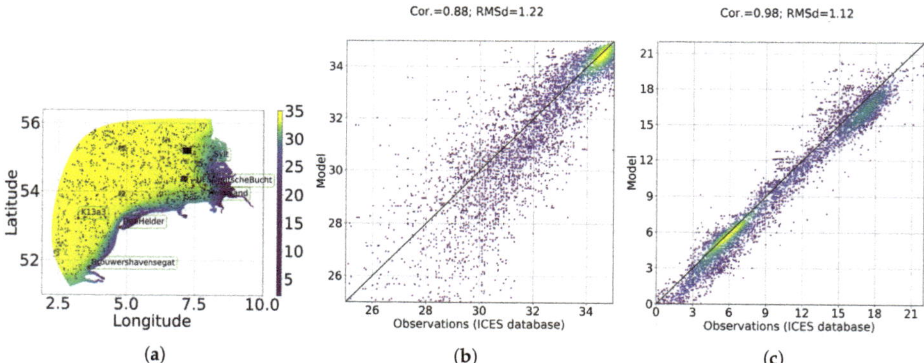

Figure 4. (**a**) Mean sea surface salinity for the years 2010–2014. The black dots indicate positions of data from the ICES database; black-with-red circles indicate positions of stations giving temperature and salinity time series with the corresponding names. (**b**,**c**) Sea surface salinity and temperature from FESOM-C (y-axis) compared one-to-one with values from the ICES database (x-axis) with corresponding correlation coefficients (Cor.) and root mean square difference (RMSd).

3.5. Time Series of Temperature and Salinity

The southern North Sea area is rich in observational data, making this area interesting for model calibration and validation. We validated the model using an automated validation system introduced in the new model version. A special model-structure output module provides output at discrete predefined stations for direct comparison to the observed time-series collected in various databases. The collection is much larger than the set of stations dealt with above; we only show comparisons for stations with interesting dynamics from which continuous measurements are available.

In Figures 5–7, a time-series of simulated temperature and salinity for the years 2010–2015 is compared with observational data from several autonomous stations. Most of the data taken from the Emodnet and COSYNA databases underwent automatic quality control. However, some data remain doubtful, for example a somewhat high temperature reading at the Sylt station during the winter of 2012–2013 (Figure 5c). Filtering and cleaning observations is beyond the scope of this paper. Nevertheless, most of the data are trustworthy and useful for direct comparison.

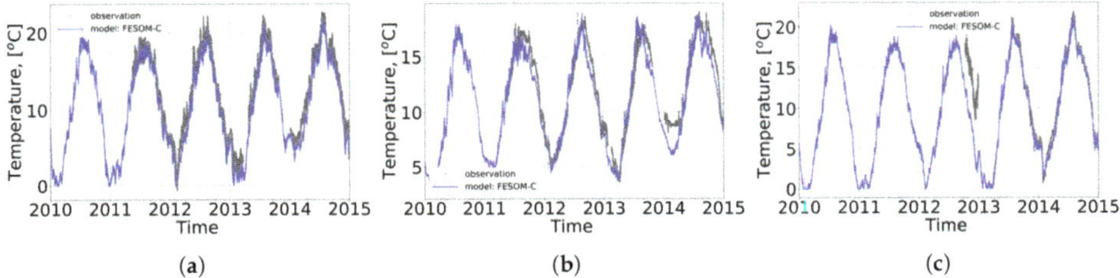

Figure 5. Observed (grey dots) and modeled (blue line) sea surface temperature at three stations: (**a**) Den Helder; (**b**) K13a3; and (**c**) Sylt. Station positions are indicated in Figure 4.

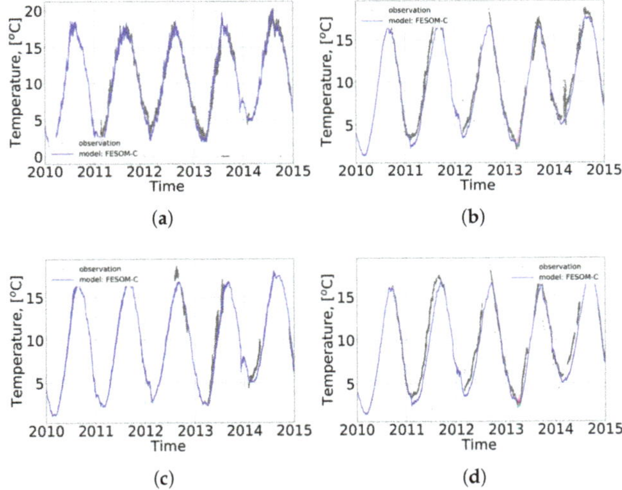

Figure 6. Comparison of observed (grey dots) and simulated (blue line) temperature at two stations: (**a,c**) the Helgoland station at 1 and 10 m depth, respectively; and (**b,d**) the UFS Deutsche Bucht station at 20 and 30 m depth, respectively.

The seasonal cycle of surface temperatures at three stations, namely K13, Den Helder, and Sylt (Figure 5), as well as at Helgoland (Figure 6a), was captured well by the model. The Den Helder station (Figure 5a) is situated on the first Wadden Sea inlet and experiences intense water exchange between the inner Wadden Sea and the open sea; in part, this explains the high temporal variability in observed and simulated temperature. The K13a3 station (Figure 5b) is situated close to the model domain's open boundary (here representing the open sea) and is less affected by coastal processes. The Sylt station (Figure 5c) is situated close to one of northern Wadden Sea bays used for various model test cases and near a salinity gradient under the influence of freshwater. The Helgoland station is characterized not only by highly variable salinity and a strong lateral salinity gradient but also by a significant bathymetric gradient. In general, the model captured observed surface temperature dynamics, including cyclical-seasonal and local effects, very well at these stations. The model generates a lower autumnal temperature at the K13a3, but there is no such effect at the other stations.

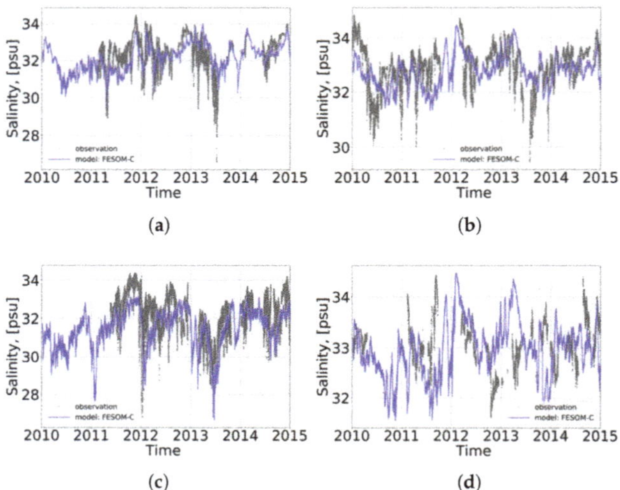

Figure 7. Comparison of observed (grey dots) and simulated (blue line) salinity at three stations: (**a**) data from the Helgoland station at 1 m depth; (**c**) data from the Brouwershavensegat station at 1 m depth; and (**b,d**) data from the UFS Deutsche Bucht station at 6 and 30 m depth, respectively.

Temperature validation for deeper layers is shown in Figure 6 for the two stations Helgoland (1 and 10 m depth) and UFS Deutsche Bucht (20 and 30 m depth). UFS Deutsche Bucht is not far from Helgoland; however, there is a deep trench leading to the open sea from the latter. Here, temperature undergoes a pronounced seasonal cycle with less seasonal variability than at the surface. The model exhibits greater deviation from observations in deep layers than at the surface. It represents observations reasonably well in general, albeit with local differences up to several degrees.

Figure 6 represents observed and simulated salinity time-series at three stations: Helgoland (1 m depth), UFS Deutsche Bucht (6 and 30 m depth), and Brouwershavensegat. The Brouwershavensegat station (1 m depth) is located near the open boundary and is affected by water from the Rhine. There are no pronounced seasonal salinity dynamics in the SeNS area ([11]). Salinity changes at these stations are determined mainly by changes in wind-driven currents and freshwater supply from rivers. Observational data variability on small time scales (of days) is generally higher than in model results; this is to be expected under the current setup, with 21 sigma layers and horizontal resolutions of up to 1 km that will not resolve river plumes and freshwater lenses properly. While the model does not track rapid changes in observed salinity, it does capture common dynamics well. Drops in salinity due to extreme flooding events such as the one in 2013 are clearly seen in observation and reproduced by the model, as shown in Figure 7a–c. The model shows a drop salinity at 30 m depth during 2013, unlike observation (see Figure 7d). This could be related to the current setup's rough vertical resolution and parameterization of vertical mixing. The most significant difference between observation and the model appears for the year 2011.

3.6. Salinity and Temperature, Ferry Lines

The near-shore area of the SeNS is characterized by strong lateral density gradients defined by the salinity gradient, mainly due to fresh water supply from rivers ([11]), and by the temperature gradient under the effect of the different temperature dynamics of shallow and deep areas. Such gradients may play a significant role in near-coastal dynamics ([34,53,54]). Temperature and salinity data collected by the Operational Systems department of the Institute of Coastal Research Helmholtz-Zentrum Geesthacht, by way of FerryBox systems installed on several ferries connecting Cuxhaven–Helgoland (CH), Büsum–Helgoland (BH), and Cuxhaven–Immingham (CI) [47,55,56], offers a way to verify the model both in near-coastal areas (BH and CH) as well as farther offshore (CI). These ships' areas of operation are indicated by the three ellipses in Figure 8, superimposed over their respective paths indicating salinity values. Raw data for analysis were taken from the COSYNA database. The CH and BH ferries both call at the island of Helgoland, where mean salinity is about 34 PSU. The Port of Cuxhaven (serving CH and CI) is situated at the western side of the mouth of the river Elbe, and the port at Büsum is about 30 km north of the Elbe estuary. Both are located in the region of freshwater influence from the Elbe, with horizontal salinity gradients up to 0.45 PSU/km and tidal amplitudes in excess of 1.5 m, with a broad wetting-and-drying area. We used data from the CI ferry between $3°$ E longitude and Cuxhaven; the rest lay outside the model domain. One axis in the figures tracks longitude; the sailing routes run mainly east–west. The CI ferry altered its route over time, as is reflected in the mean values (Figure 9).

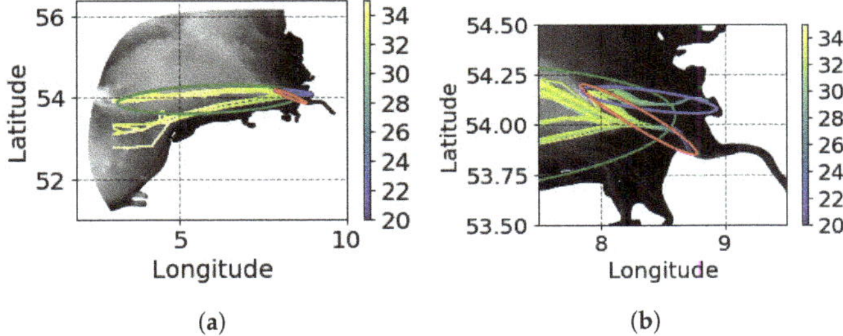

(a)   (b)

Figure 8. Sea-surface salinity measured by three ferry lines across their paths used for comparison in Figures 10–12. Greyscale background shows bathymetry. The color scatter plots show observed salinity for the period 2010–2014. The three ferries' areas of operation are indicated by ellipses: (**b**) red for Cuxhaven–Helgoland (CH), blue for Büsum–Helgoland (BH), and (**a**) green for Cuxhaven–Immingham (CI).

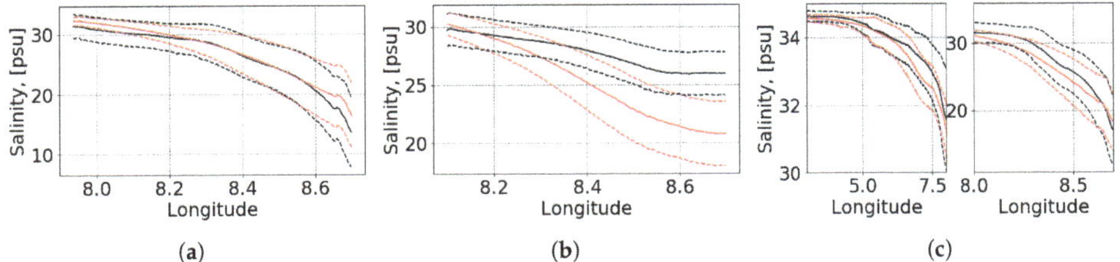

(a)   (b)   (c)

Figure 9. Mean sea-surface salinity from the three ferry lines (black lines) and corresponding model values (red lines). Mean values are shown by solid lines; mean values ± one standard deviation are shown by dashed lines. Ferries: (**a**) Cuxhaven–Helgoland; (**b**) Büsum–Helgoland; and (**c**) Cuxhaven–Immingham.

The data sampling rate varies from 10 to 50 s and depends on time and route; the distance between data samples thus depends on a ship's speed and the sampling rate. The distance between neighboring measurements across the dataset used here varies from 80 to 400 m. Both the spatial (80 m) and temporal (10 s) resolutions are much higher than in the model (1000 m in coastal areas and 60 s, respectively) and especially higher than in the three-dimensional model output (with the same spatial resolution but 1.5-hourly mean values). The number of model-3D-output snapshots is sharply limited by the space required to store the output data and the performance of long-term memory (HDD and SSD), and could not be significantly changed. Observed data are also scattered across time and space, which limits the ability to save model output at exact times and positions as doing so would mean storing more than 2 million scattered data points that are not yet realized in the model. Subsequent versions of the model will include this feature. The distinction between the sampling rate and model output introduces a discrepancy into any direct comparison. A temperature difference of 0.2–0.4 °C arises due to the time shift. Comparing coarse horizontal model resolution to observation leads to a situation in which several observational points correspond to a single model point. Due to salinity gradients, and accounting for tidal currents of up to 2 m/s, errors in salinity can approach ±2 PSU. The more transects are sailed, the less the mean values will differ due to random errors such as in spatial and temporal resolution; however, errors in deviation will remain. Greater differences are expected in near-coastal areas because the water masses there vary much more. Salinity goes from up to 35 PSU near Helgoland to brackish near land, with salinities between 5 and 10 PSU. Direct comparison among different instruments (FerryBox, water sample analyses, OSTIA satellite data, and MARNET stations) were provided by Haller et al. [33], Grayek et al. [57]. Haller and colleagues [33] reported an error of 0.79 RMS in salinity between FerryBox data and laboratory water sample analyses. Grayek and colleagues [57] found that temperature differences between FerryBox and MARNET station readings can be the result of heating of the measured water in the ferry.

The difference in the density of water masses is known to determine the dynamics of the baroclinic system. Similar to Gräwe et al. [11], in Figures 10 and 11, we give temperature and salinity anomalies for the three ferry lines. Apparently, both density constituents (for salinity, see Figure 10; and, for temperature, see Figure 11) play a significant role in the formation of the density gradient (Figure 12). This comparison allows us to evaluate the potential source of error in the model. Anomalies were calculated separately for each ferry section, eliminating seasonal differences. However, comparing only the anomalies in the model does not inform us about absolute values. The statistics for this comparison are shown in Table 3 and Figures 9 and 13 for absolute values and anomalies.

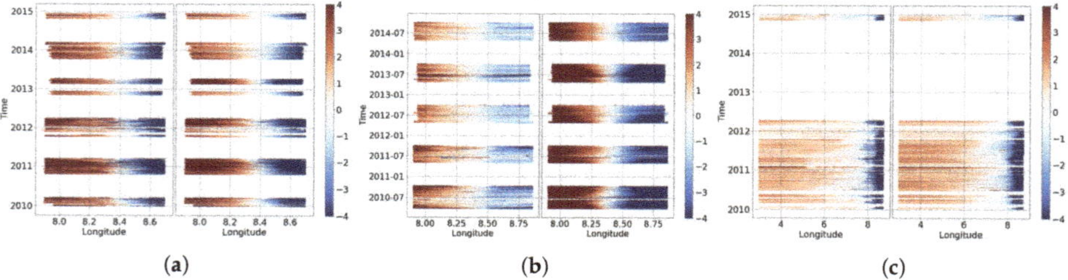

Figure 10. Salinity anomalies from the three ferry lines and corresponding model results. The observational data are on the left and the corresponding model results are on the right of each panel. Ferries: (**a**) Cuxhaven–Helgoland; (**b**) Büsum–Helgoland; and (**c**) Cuxhaven–Immingham.

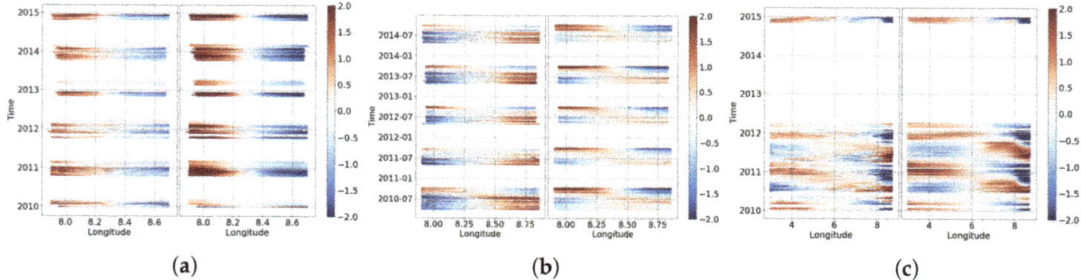

Figure 11. Temperature anomalies from the three ferry lines and corresponding model results. The observational data are on the left and the corresponding model results are on the right of each panel. Ferries: (**a**) Cuxhaven–Helgoland; (**b**) Büsum–Helgoland; and (**c**) Cuxhaven–Immingham.

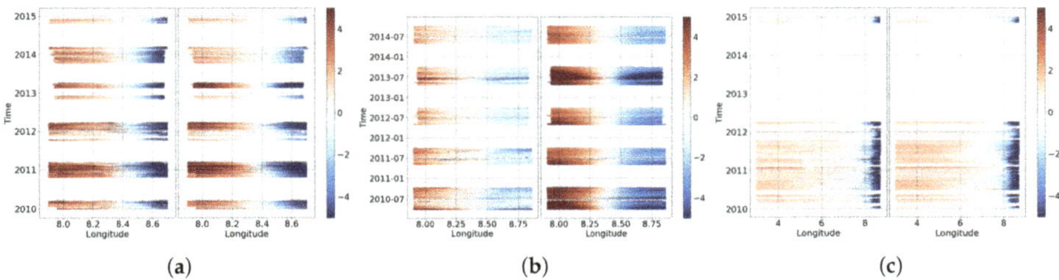

Figure 12. Density anomalies from the three ferry lines and corresponding model results. The observational data are on the left and the corresponding model results are on the right of each panel. Ferries: (**a**) Cuxhaven–Helgoland; (**b**) Büsum–Helgoland; and (**c**) Cuxhaven–Immingham.

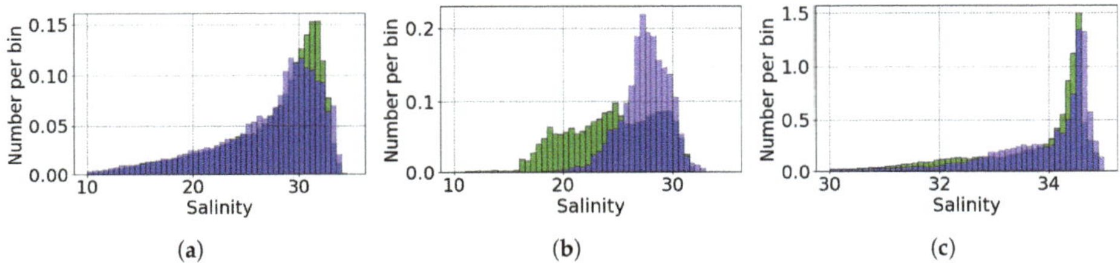

Figure 13. Distribution of salinity in measurements from the three ferry lines (blue) and the model (green). Ferries: (**a**) Cuxhaven–Helgoland; (**b**) Büsum–Helgoland; and (**c**) Cuxhaven–Immingham.

Table 3. Comparison of salinity (S), temperature (T), and density (ρ) between FerryBox and FESOM-C. RMSE, Root Mean Square Error of anomalies. The numbers in brackets are based on absolute values.

FerryBox (Operation Area)	Number of Measurements	Number of Transects	S, RMSE	T, RMSE	ρ, RMSE
Cuxhaven–Helgoland	≈234,000	583	1.7 (2.2)	0.8 (2.7)	1.3 (1.8)
Büsum–Helgoland	≈502,000	1171	2.3 (3.8)	0.7 (2.9)	1.7 (2.5)
Cuxhaven–Immingham	≈1,552,000	350	0.9 (1.0)	0.6 (1.2)	0.7 (0.7)

Basic statistics for the available data and comparison with the model are shown in Table 3. The total number of measured points from over 2000 ferry transects is more than 2,000,000. Data from the CI and BH ferry lines cover the whole modeled period of 2010–2015; the gaps (the white area in the figures) are seasonal, reflecting winter (BH) or summer (CH). The CI ferry line has fewer gaps for 2010–2012. The second part of the simulated period is not covered by data from the CI ferry except for some data from the end of 2014. The CI ferry line traces a much longer transect compared to the other two and mostly represents water masses with a high salinity of more than 34 PSU (see Figure 13, blue bars). The BH and CH ferry lines mostly sail near the fresh water influence area of the river Elbe, where salinity varies 10–35 PSU for CH and 20–35 PSU for BH. The RMS error of anomalies is smaller in general than the corresponding RMSE of absolute values shown in brackets. Gräw and colleagues [11] compared results of the well-established GETM model with BH ferry data for the 2009–2011 period. The horizontal resolution of the GETM grid presented by Gräwe et al. [11] is 200 m, that is, five times finer than the present setup. Gräwe et al. [11] reported RMSEs for salinity of 1.15 PSU, 0.64 °C for temperature, and 1.32 kg/m^3 for density. Haller and colleagues [33] compared the results of two three-dimensional hydrodynamic models, BSHcmod and AMM7, with the CI ferry line the years 2009–2012. BSHcmod v4 is a three-dimensional operational model with a two-way nesting approach and a finest resolution of 900 m. AMM7 is a one-way nesting, operational model based on the NEMO model that includes assimilation of in-situ observations at a 7-km horizontal resolution.

Haller et al. [33] calculated salinity RMSEs of 0.68 and 1.1 PSU and temperature RMSEs of 0.68 and 0.44 °C for the BSHcmod v4 and AMM7 models, respectively.

The results of FESOM-C are within common statistics and clearly show better agreement in the open sea. However, comparison with the BH ferry data indicates a problematic area.

The anomalies of temperature, salinity, and density are shown in Figures 10–12, respectively, for the three ferry lines CH, BH and CI. The seasonal cycle is easily visible in the shift in positive and negative temperature anomalies from summer to winter. Such a shift (both in time and position) is reproduced well by the FESOM-C model on a small spatial scale of about 60 km (Figure 11b) as well as on longer spatial scales of 300 km (Figure 11c). Some of the small features, such as surface warming at 8° E during the autumn of 2014 (see Figure 11c), are also captured reasonably well by the model. The RMS difference for the CI ferry line is close to the comparison made with data from the ICES database (Figure 4). The difference between the mean observed and the modeled temperature varied from 0.4 to 0.8 °C and increased with rising temperatures. The temperature difference increases significantly towards the coast in shallow waters. Several possible reasons for this mismatch are worth mentioning; although their importance is known, the current setup still does not implement them. The depth to which short-wave solar radiation penetrates varies significantly in this region and depends heavily on suspended matter. There is a well-known steep gradient in suspended matter along this region's coast. The lack of feedback with the atmospheric model used at the upper-boundary ocean model is an issue too, since the atmospheric model assumes surface-water temperatures different from those simulated by FESOM-C. That leads, in turn, to incorrect long-wave radiation ([58]). Inaccuracy in sea-bed albedo, together with wave effects, might also modify surface heat flux. These shortcomings of boundary condition formulation for tracer equations will be taken into account in the next version of the model.

Unlike temperature, salinity, and density do not show a pronounced seasonal cycle. However, salinity exhibits steep, more pronounced offshore gradients. The model reproduces salinity and density anomalies reasonably well, as it does with temperature. Most of the differences appear in shallow areas with maximum lateral gradients. Variability in salinity is generally close to what is observed, except for one area near Helgoland (the western port of the CH and BH ferries; see Figure 9a,b). Here, the model's standard deviation is about two times smaller than observations while mean salinity is close to observations. The temporal dynamics are captured reasonably well by the model at the Helgoland station (see the salinity time-series in Figure 7a). Simulated mean salinity starts to deviate significantly from observation towards the coast in the area north of the mouth of the Elbe River (the BH ferry line). The model reproduces observed salinity dynamics west of the Elbe (the CH ferry line) well.

Several factors may explain the model's deviation in the area of the BH ferry line. The tidal wave propagates from west to east up to the Elbe River and then northward along the coast. The tidal dynamics near Cuxhaven (the eastern port of the CH ferry) is reproduced well by the model. However, north of the Elbe River, the model shows a significantly lower amplitude for the M2 tidal wave (see Figure 2, Station 12). Stations 14 and 15 in the same figure, lying further north, do not show such a significant deviation. Areas of wetting and drying play a significant role in this region's dynamics. A significant near-coastal area falls dry during low tide. Errors in the model's bathymetry—the area north of the Elbe River is deeper—and coarse resolution in the wetting-and-drying areas, together with uncertainties in bottom drag, lead to higher Elbe-water transport along the coast. However, most likely, it should rather propagate in a north-westerly direction, towards Helgoland. A sensitivity study with higher resolution near the coast and a more accurate representation of the bathymetry improved both barotropic and baroclinic model dynamics in this area. The results of these studies are partly presented in Section 4.

3.7. Vertical Structure, Gliders

For the 2010–2014 period, the Operational Systems department of the Institute of Coastal Research Helmholtz-Zentrum Geesthacht performed several surveys, with two gliders called "Sebastian" and "Amadeus", in the area of Helgoland and towards the open sea. Data from these surveys (which cover an area not reflected in the FerryBox data; see Section 3.6) are available in the COSYNA database. With high vertical and special resolutions, these data offer a unique opportunity to validate the model. Modeled salinity and temperature are compared with glider data in Figures 14 and 15, respectively, in which bathymetry with glider path is shown at right, model data at top left, and glider data at bottom left of the panels showing the two surveys. Time of measurement is shown on the map by color, from red (beginning of survey) to yellow (end of the survey); this corresponds to the time axis on the plots. Every glider measurement is shown by a colored circle. The model output was interpolated to match the glider data in time and space by the "nearest neighbor" method. For every glider data point, one profile corresponding to the time of measurement is extracted from the model output, to exclude duplicating profiles. The "nearest neighbor" method introduces some inaccuracy and moreover allows only point-by-point comparison between the model and observations. The measured data points total about 60,000 (winter transect) and 26,000 (summer transect), vastly fewer than the FerryBox data, but are distributed in 3D space and so require three-dimensional interpolation. 3D linear interpolation of the model results on an unstructured mesh, as was done for the 2D FerryBox data, was dispensed with as time and memory consuming. Furthermore, the 3D model output that formed the basis for comparison was saved at 1.5-h intervals. The time resolution of the glider data is about 1 min, so the horizontal resolution of the observed data is thus significantly higher than in the model. The 1.5-h time resolution of the model output introduces an additional discrepancy as against observation due to tidal phase shifts. The water masses may have shifted about 5 km horizontally, given a current of 1 m/s. These uncertainties in data comparison should be taken into account during future analysis. However, the resulting figures are nevertheless useful to point out the advantages and disadvantages of the model.

The results of two glider transects from the winter of 2011 and the summer of 2013 are shown in Figure 14 (salinity) and Figure 15 (temperature). The 2011 winter glider path was from near Helgoland (9 February 2011) northwest for about 60 km and back (22 February, 2011). Quality checked salinity data are available only for the second part of the transect, starting on 15 February, 2011. Both the temperature and the salinity figures show a well-mixed water column in observations and model results. Shortly before and during the winter transect, winds were strong as compared to the summer transect. Mean wind speed was about 8 m/s (and up to 16 m/s) with mean air temperatures around 0 °C near Helgoland. This explains the mixed water column and slight overcooling at the surface in the model. Horizontal temperature (lower temperature near Helgoland) and salinity (increase salinity towards the open sea) gradients are reproduced reasonably well by the model. Observations show wavy variability in both the temperature and salinity profiles, with a period similar to the M2 tidal wave, which is probably the effect of a tidal dynamic. The model captures a similar dynamic well. Unlike the winter situation, the summer transect (see Figures 14b and 15b) shows strong vertical stratification in the temperature fields. Mean wind speed on the summer transect was 5.5 m/s (and up to 12 m/s), mean air temperature about 18 °C, and short-wave solar radiation was about four times higher than in winter, determining the extant strong thermocline. While the model shows similar sea-surface and near-bottom temperature dynamics, it does not capture the sharp vertical gradient. The simulated temperature and salinity profiles are smoother than the observed ones. The model does not resolve surface freshwater plumes near Helgoland. Vertical gradients in modeled salinity profiles are much less pronounced, and smoothed, compared to the vertical temperature structure. The current setup has 21 sigma layers, which may not be enough to capture steep vertical gradients. A sensitivity test with more layers shows some improvements in model results. Improvements in vertical turbulence schemes are also required.

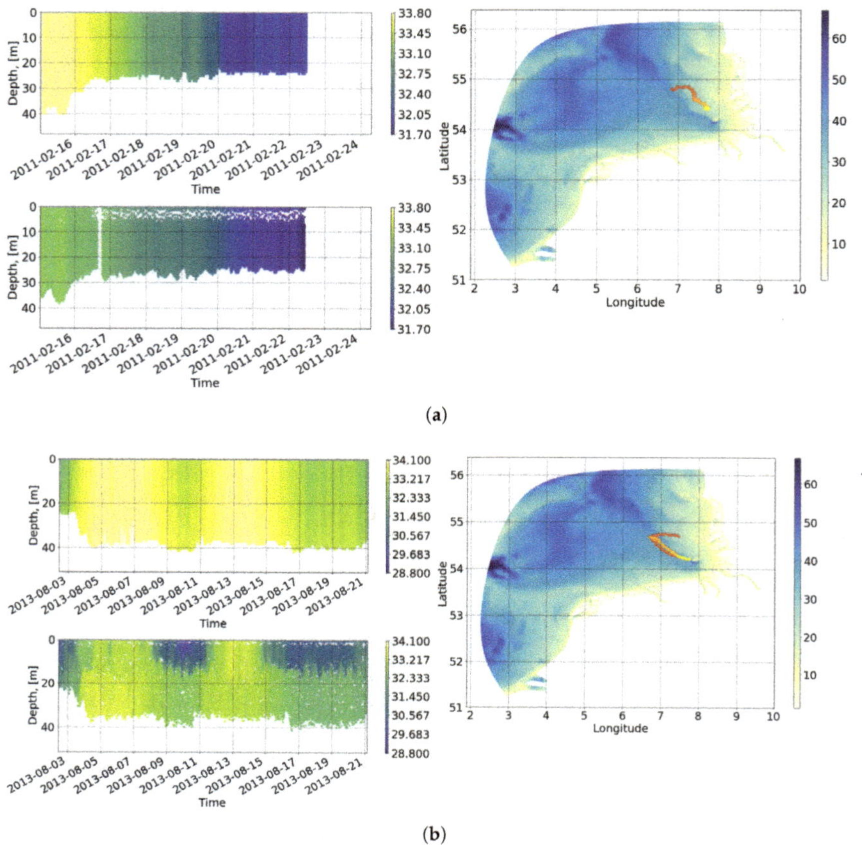

Figure 14. Comparison of model salinity with observed data from gliders: (**a**) February 2011; (**b**) August 2013. Colored circles on bathymetry maps (right) show glider paths; color indicates position times from red (beginning of the survey) to yellow (end of the survey). The filled contours at upper left are model results (21 sigma levels, 2010–2014 run). The scatter data at bottom left are from gliders (COSYNA database ([51])). The glider data time resolution is about 1 min, or 20 m horizontal resolution. The spatial resolution of the model is between 1 and 4 km.

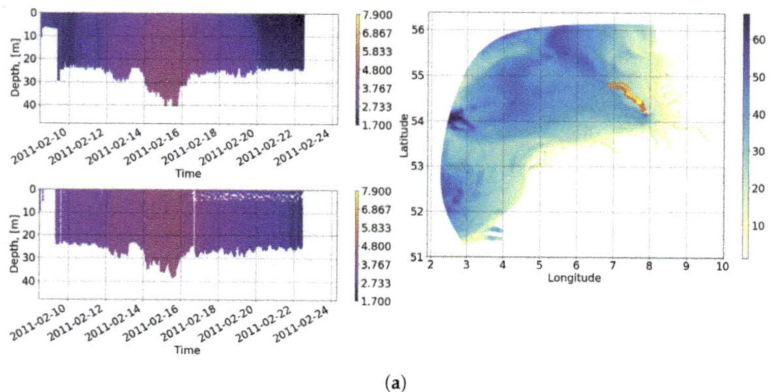

(**a**)

Figure 15. *Cont.*

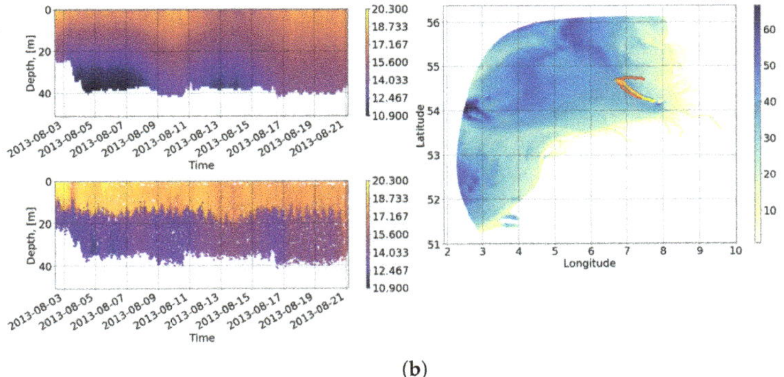

(b)

Figure 15. Comparison of model temperature with observed data from gliders: (**a**) February 2011;
(**b**) August 2013. Colored circles on the bathymetry maps (at right) show glider paths; colors indicate
position times from red (beginning of the survey) to yellow (end of the survey). The filled contours at
upper left are model results (21 sigma levels, 2010–2014 run). The scatter data at bottom left are from
gliders (COSYNA database [51]). The glider data time resolution is about 1 min, or 20 m horizontal
resolution. The spatial resolution of the model is between 1 and 4 km.

4. Discussion, Mesh Resolution

Comparing the model results with observations, we find the largest discrepancy in shallow areas
near the coast, especially in wetting-and-drying areas. Previous work, as well as sensitivity studies
with the current model, have shown that refining the horizontal resolution of the mesh significantly
improves model results. At the same time, it is not always obvious what horizontal resolution different
regions require. To find the optimal resolution, we performed several sensitivity simulations.

Solution Convergence on Different Meshes

One of the important stages during preparation is selecting the optimal mesh resolution for the
modeled region. We understand optimal mesh resolution as a compromise between computational
efficiency and the quality of the simulated dynamics. A preliminary calculation on the sequence of
meshes allows us to estimate the convergence of numerical solutions. We constructed three different
meshes for our experiments. All were generated by the Gmsh mesh generator ([39]). The first
one (m8) has a spatial resolution of between 4 and 1 km and 43,318 vertices. The resolution of
the second mesh (m5) varies from 2.2 to 550 m and has 134,858 vertices. The third mesh (m3) has
minimum and maximum cell sizes of 250 m and 1.6 km, respectively, and 235,283 vertices. All meshes
have 21 non-uniform sigma layers in the vertical direction (refined near the surface and bottom).
The wetting/drying option was turned on. To test the code's sensitivity to the mesh resolution,
we computed barotropic, tidally driven circulation in the SeNS for two atmospheric scenarios: one
without a strong wind effect for the full tidal period ("weak wind"), and the other with a "strong wind"
component. Discrete sea-surface height (SSH) values and two components of velocity (u, v) were
accumulated in the mesh vertices, and then these values were linearly interpolated onto the vertices of
the coarse mesh (m8). Next, we determined the values of the different $\delta\zeta$, δu, and δv among solutions
on the meshes. We performed the comparison for the full M2 tidal cycle, as shown in Figure 16, which
presents the histograms of the differences.

For the solutions on the m8 and m5 meshes for the elevation values (the tidal wave maximum
exceeds 3.5 m), only 30.6% of the points agree within a range of ±1 cm for the "strong wind"
experiment; 31.7% agree in the "weak wind" scenario (Figure 16(top)a,b). For the ±2 cm interval,
the proportion of points with the same solution more than doubles, while exceeding 67% for both
scenarios. A similar situation occurs when we compare horizontal velocity components on these
two meshes (Figure 16(bottom)a,b). In the range of ±1 cm/s, the number of identical points for the

u-component is approximately 43% in both scenarios but slightly exceeds 48% for the *v*-component (the maximum horizontal velocity is about 280 cm/s). For a ±2 cm/s interval, the number of identical solutions is about 80% for both velocity components.

Comparing the more detailed m5 and m3 meshes, we see a significant improvement in convergence of the results, with the exception of the sea-surface height comparison in a "weak wind" scenario (Figure 16(top)c,d). Here, the number of points falling within the ±1 cm limit is only 27.9%; in a "strong wind" experiment, the number of such matches is almost twice as large (55.6%). Increasing the match interval up to ±2 cm, the number of matches increases significantly, to 73.3% for a "weak wind" and 85.1% for a "strong wind" scenario. A significant improvement in convergence of the results is also apparent in the horizontal velocity field (Figure 16(bottom)c,d). Thus, in the range of ±1 cm/s, the number of matches in the "weak wind" experiment is 55% for the transverse velocity component and 62% for the longitudinal velocity component. In the "strong wind" case, the convergence of results is greatly improved, exceeding 70% for both components of the velocity vector. In the range of ±2 cm/s, the convergence of solutions tops 90% for the velocity components of the two experiments.

As is apparent from the foregoing results (see Figure 16), the experimental "strong wind" simulations negate the poor spatial resolution somewhat, especially in the shallow-water zone where mesh resolution errors are highest. Either onshore winds increase the thickness of the water, or offshore winds drain the shallow-water zones faster; the result is smoother errors on meshes of different resolutions.

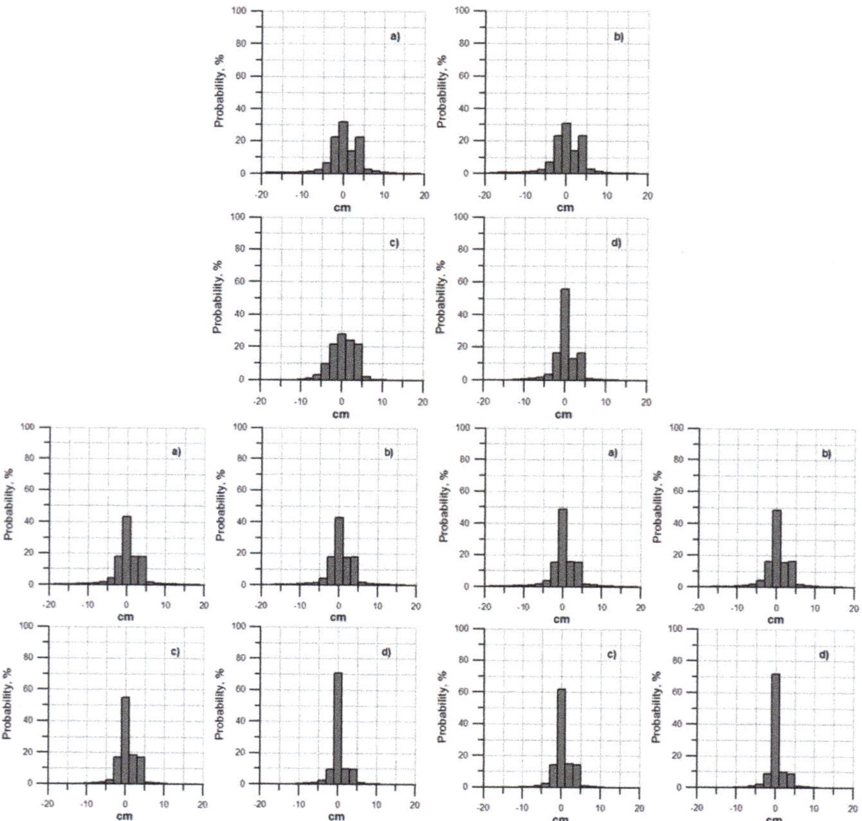

Figure 16. Histograms of the difference between solutions at different mesh resolutions: (top) sea-surface elevation; (bottom left) the *u*-component of velocity; and (bottom right) the *v*-component of velocity. (**a**) Meshes m8 and m5 ("weak wind"); (**b**) Meshes m8 and m5 ("strong wind"); (**c**) Meshes m5 and m3 ("weak wind"); and (**d**) Meshes m5 and m3 ("strong wind").

As can be seen from the results of the spatial comparison (Figure 17), the maximum difference falls in zones of drying and of minimum depth. In detailing the coastal zone, the wetting-and-drying processes differ, to an extent, from the solution on coarser meshes. The difference between the solutions on the coarse and middle-resolution meshes in the "weak wind" case reveals a certain difference in the solutions for the deep-water part of the region precisely in the amphidromic zone of the M2 wave (Figure 17a). The difference never exceeds 0.2 cm, meaning there is a slight shift of the amphidromic point for the solutions on the two meshes. The solutions on the intermediate and on the most detailed meshes do not show this difference in the amphidromic zone (Figure 17b).

A different situation arises in the "strong wind" scenario: Two zones of maximum difference appear in a deep-water part of the SeNS. Their locations stem from the region's bathymetric features, namely a small underwater sill. At the same time, a coarse-mesh solution produces somewhat underestimated results in elevation at localized areas not exceeding 0.25 cm (Figure 17c). There is practically no difference between the solutions on the intermediate and the most detailed meshes (see Figure 17d). This analysis shows that, for model simulations, the m5 mesh will yield solutions of optimal quality but also that a coarser mesh (m8) solution will not introduce significant errors in the model results.

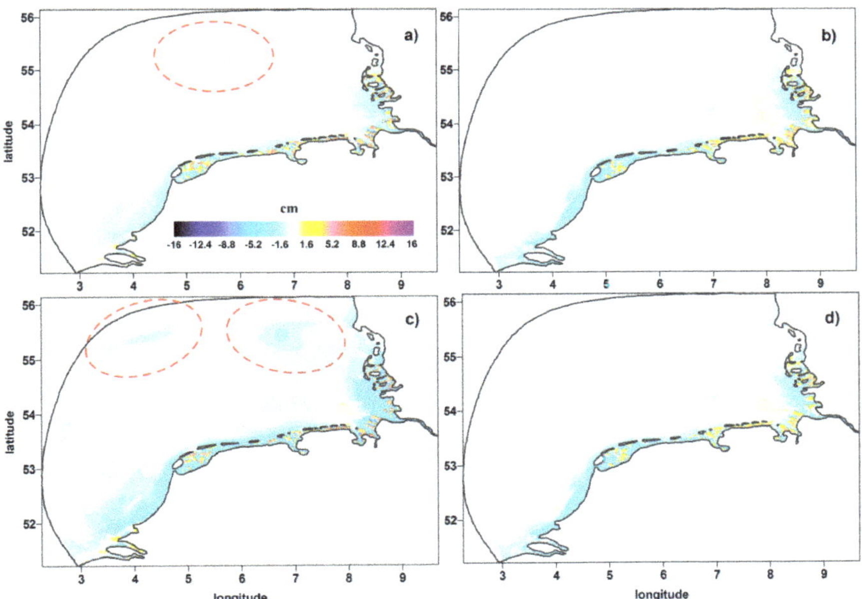

Figure 17. The spatial difference for SSH between solutions on different meshes with different atmospheric scenarios: (**a**) at $\delta\zeta^{m8} - \delta\zeta^{m5}$; (**b**) at $\delta\zeta^{m5} - \delta\zeta^{m3}$ in "weak wind"; (**c**) at $\delta\zeta^{m8} - \delta\zeta^{m5}$; and (**d**) at $\delta\zeta^{m5} - \delta\zeta^{m3}$ in "strong wind". Red ellipses denote zones of maximum differences in the deep-sea part of the North Sea.

5. Conclusions

The first fully realistic, three-dimensional, multi-year baroclinic hindcast simulations were performed with the newly developed FESOM-C model and were comprehensively validated in the area of the southeastern North Sea. The FESOM-C model, developed mainly for the coastal region, employs mixed unstructured-mesh methods ([25,27]) and a finite-volume discretization. FESOM-C is a fully resolved three-dimensional model based on primitive equations for momentum, continuity, and density constituents ([25]). Well-developed modules for the open boundary, the upper boundary (interaction with the atmosphere), rivers, output, and postprocessing allow this model to be used for realistic simulations. The present work uses hybrid meshes that combine quadrilaterals and a small number of triangles. Such meshes support zooming-in to an area of interest (in this case, for study of the Wadden Sea and its estuaries) even as mesh resolution towards the open sea coarsens significantly. This variable horizontal resolution enables a more efficient use of computational resources while refinement in shallow areas resolves important small-scale process (such as wetting and drying, sub-mesoscale eddies, and the dynamics of steep coastal fronts). Using meshes composed mostly of quadrilateral cells allows a significant increase in calculation rates. Proper representation of physical processes, both near shore and in the open sea, makes it possible for the model to represent their dynamics.

Overall, our validation shows that the FESOM-C model reproduces the physical dynamics of the southeastern North Sea reasonably well.

The FESOM-C model reproduces the tidal dynamics, one of the most important dynamic components of the area in question, well (on tidal dynamics, see Section 3.1). It captures the amplitudes and phases of the main tidal harmonics well compared to other solutions, and modeled sea surface height for the years 2010–2015 agrees well with observations. The model also reproduces mean and seasonal horizontal and temporal distributions of temperature and salinity reasonably well (see Sections 3.4 and 3.5). An analysis of temperature, salinity, and density readings from three FerryBoxes shows that the model can reproduce the water mass characteristics of the open sea very well, with somewhat large (but explicable) inaccuracies in the coastal zones.

An analysis of a comparison between glider data and simulated three-dimensional temperature and salinity yields estimates that allow us to evaluate the model's skill at modeling the southeastern North Sea. The vertical temperature and salinity distribution indicates that the vertical turbulence scheme needs improvement.

New developments in model output and postprocessing methods allow model results to be validated against observations on unstructured mixed meshes. Detailed validation includes various observational datasets from different autonomous instruments, such as FerryBoxes, gliders, and buoys, which are spread across time and space. The present work uses data with much higher local spatial and temporal resolutions (up to several seconds and meters) than most models, illustrating high natural variability in the coastal area.

The fact that horizontal salinity and temperature gradients as well as frontal dynamics are so well captured demonstrates that the cell-vertex (finite volumes) discretization method, used here with hybrid meshes, is capable of realistic application in this region, where the dynamics of steep gradients are a crucial and quite often difficult modeling issue. The present work should allay skepticism about unstructured-mesh coastal models as being "too dissipative" in character. In fact, they are on par with the more traditional models formulated on structured rectangular grids. The combination of coarse resolution towards the open boundary with a sharper focus on the coastal area of the Wadden Sea region offers a unique opportunity to simulate a region of interest at relatively high resolution with significantly reduced computational cost. The final computation was performed with only 24 CPUs using OpenMP parallelization. Such resources nowadays are comparable to state-of-the-art laptops and smartphones and make simulations of complicated 3D realistic cases possible without operational difficulties and expensive supercomputers. The relatively small (in terms of horizontal size and resolution) mesh used in the current setup could be extended to a larger area without significant impediments. Increasing mesh resolution by a factor of 4 leads to the number of nodes increased by a factor of 16 for a regular rectangular grid. This changes depending on the shape of the calculated domain. There is no difference between the structured and unstructured meshes in the case of the open ocean. In the proposed work, similar increase in the resolution from robust mesh (m8) to fine mesh (m3) leads to an increase in the number of nodes by a factor of 5.5. The scalability of the FESOM2 model was recently studied by Koldunov et al. [59]. In [59], the authors demonstrated that the FESOM2 model is competitive tools for a high-resolution climate modeling. In turn, FESOM-C uses the mixed (predominantly quadrilateral) meshes. Approach based on mixed meshes improves total performance and accuracy of the model in contrast to triangular meshes [27], for quadrilateral cells involving fewer edges than triangular cells. Newly developed methods for postprocessing on unstructured meshes with a rapidly growing community significantly improved the overall performance of such modeling, from preparing the setup through to final plotting. Without the problems related to one- or two-way nesting and with significantly reduced problems at the open boundary (because the meshes are compatible with global models), FESOM-C has become a realistic substitute for existing models when it comes to simulating on regional scales, from particular events to climate simulations.

6. Code Availability

The version of FESOM-C v.2 used to carry out the simulations reported here can be accessed from https://doi.org/10.5281/zenodo.2085177. The datasets needed for running this setup are available from the corresponding author upon a reasonable request.

Author Contributions: I.K., A.A., and V.F. designed experiments. I.K. setup and carried out the experiments. I.K. analyzed and visualized the model as well as the observed data. I.K. wrote the paper with the support of A.A., V.F., and S.D. A.A. analyzed and wrote up the results in the section on "solution convergence on different meshes". A.A. developed the FESOM-C model with the support of I.K., V.F., and S.D. S.H. and N.R. carried out code optimization and parallelization. I.K., A.A., V.F., and S.D. contributed with discussions of the results. K.H.W. helped supervise the project. All authors discussed the results and commented on the paper at all stages. All authors have read and agreed to the published version of the manuscript.

Funding: We acknowledge support by the Open Access Publication Funds of Alfred-Wegener-Institut Helmholtz-Zentrum für Polar- und Meeresforschung. This research was partly funded by the state assignment of FASO Russia (theme 0149-2019-0015).

Acknowledgments: We would like to thank Semjon Schimanke from SMHI for help in preparing and providing the atmospheric forcing. We would also like to thank the teams behind the COSYNA, ICES, and EMODnet databases for data gathering and maintenance. We thank the administrators of the AWI cluster Ollie, where the final simulations took place, for their continual support and patience. We moreover thank the many people from the Institute of Coastal Research Helmholtz-Zentrum Geesthacht (HZG), where IK began preparing the current manuscript, for their great cooperation and useful discussions, in particular: Holger Brix, for general support at the early stages of preparing the current manuscript; Yoana G. Voynova and Onur Kerimoglu, for discussing the FerryBox data; and Onur Kerimoglu again, for sharing river discharge data. We thank reviewers who have contributed their time and expertise to review manuscript.

Conflicts of Interest: The authors declare no conflict of interest.

References

1. Danilov, S.; Sidorenko, D.; Wang, Q.; Jung, T. The finite-volume sea ice-Ocean model (FESOM2). *Geosci. Model Dev.* **2017**, *10*, 765–789. [CrossRef]

2. Madec, G.; The NEMO System Team. *NEMO Ocean Engine, Version 3.6 Stable*; Technical Report 27; note du Pôle de modélisation de l'Institut Pierre-Simon Laplace No 27; IPSL, 2015; Available online: http://www.nemo-ocean.eu/; .

3. Shchepetkin, A.F.; McWilliams, J.C. The regional oceanic modeling system (ROMS): A split-explicit, free-surface, topography-following-coordinate oceanic model. *Ocean Model.* **2005**, *9*, 347–404. [CrossRef]

4. Griffies, S.M. *Elements of the Modular Ocean Model (MOM): 2012 Release*; Technical Report C; NOAA/Geophysical Fluid Dynamics Laboratory: Princeton, NJ, USA, 2012.

5. Burchard, H.; Bolding, K. *Getm—A General Estuarine Transport Model. Scientific Documentation*; Technical report EUR 20253 en; European Commission, Joint Research Centre, Institute for Environment and Sustainability: Copenhagen, Denmark, 2002; p. 164.

6. Chen, C.; Huang, H.; Beardsley, R.C.; Liu, H.; Xu, Q.; Cowles, G. A finite volume numerical approach for coastal ocean circulation studies: Comparisons with finite difference models. *J. Geophys. Res. Ocean.* **2007**, *112*. [CrossRef]

7. Izquierdo, A.; Mikolajewicz, U. The role of tides in the spreading of Mediterranean Outflow waters along the southwestern Iberian margin. *Ocean Model.* **2019**, *133*, 27–43. [CrossRef]

8. Murray, S.P.; Arief, D. Throughflow into the Indian Ocean through the Lombok Strait, January 1985–January 1986. *Nature* **1988**, *333*, 444–447. [CrossRef]

9. Marchesiello, P.; McWilliams, J.C.; Shchepetkin, A. Open boundary conditions for long-term integration of regional oceanic models. *Ocean Model.* **2001**, *3*, 1–20. [CrossRef]

10. Marsaleix, P.; Auclair, F.; Estournel, C. Considerations on open boundary conditions for regional and coastal ocean models. *J. Atmos. Ocean. Technol.* **2006**, *23*, 1604–1613. [CrossRef]

11. Gräwe, U.; Flöser, G.; Gerkema, T.; Duran-Matute, M.; Badewien, T.H.; Schulz, E.; Burchard, H. A numerical model for the entire Wadden Sea: Skill assessment and analysis of hydrodynamics. *J. Geophys. Res. Ocean.* **2016**, *121*, 5231–5251. [CrossRef]

12. Weisse, R.; Bisling, P.; Gaslikova, L.; Geyer, B.; Groll, N.; Hortamani, M.; Matthias, V.; Maneke, M.; Meinke, I.; Meyer, E.M.; et al. Climate services for marine applications in Europe. *Earth Perspect.* **2015**, *2*. [CrossRef]

13. Debreu, L.; Marchesiello, P.; Penven, P.; Cambon, G. Two-way nesting in split-explicit ocean models: Algorithms, implementation and validation. *Ocean Model.* **2012**, *49–50*, 1–21. [CrossRef]

14. Herzfeld, M.; Rizwi, F. A two-way nesting framework for ocean models. *Environ. Model. Softw.* **2019**, *117*, 200–213. [CrossRef]

15. Sheng, J.; Greatbatch, R.; Zhai, X.; Tang, L. A new two-way nesting technique for ocean modeling based on the smoothed semi-prognostic method. *Ocean Dyn.* **2005**, *55*, 162–177. [CrossRef]

16. Chen, C.; Liu, H.; Beardsley, R.C. An unstructured grid, finite-volume, three-dimensional, primitive equations ocean model: Application to coastal ocean and estuaries. *J. Atmos. Ocean. Technol.* **2003**, *20*, 159–186.<0159:AUGFVT>2.0.CO;2. [CrossRef]

17. Qi, J.; Chen, C.; Beardsley, R.C. FVCOM one-way and two-way nesting using ESMF: Development and validation. *Ocean Model.* **2018**, *124*, 94–110. [CrossRef]

18. Zhang, Y.J.; Ye, F.; Stanev, E.V.; Grashorn, S. Seamless cross-scale modeling with SCHISM. *Ocean Model.* **2016**, *102*, 64–81. [CrossRef]

19. Zhang, Y.; Baptista, A.M. SELFE: A semi-implicit Eulerian-Lagrangian finite-element model for cross-scale ocean circulation. *Ocean Model.* **2008**, *21*, 71–96. [CrossRef]

20. Burchard, H.; Schuttelaars, H.M.; Ralston, D.K. Sediment Trapping in Estuaries. *Annu. Rev. Mar. Sci.* **2017**, *10*, 371–395. [CrossRef]

21. Fransner, F.; Nycander, J.; Mörth, C.M.; Humborg, C.; Markus Meier, H.E.; Hordoir, R.; Gustafsson, E.; Deutsch, B. Tracing terrestrial DOC in the Baltic Sea—A 3-D model study. *Glob. Biogeochem. Cycles* **2016**, *30*, 134–148. [CrossRef]

22. Hordoir, R.; Axell, L.; Höglund, A.; Dieterich, C.; Fransner, F.; Gröger, M.; Liu, Y.; Pemberton, P.; Schimanke, S.; Andersson, H.; et al. Nemo-Nordic 1.0: A NEMO-based ocean model for the Baltic and North seas—Research and operational applications. *Geosci. Model Dev.* **2019**, *12*, 363–386. [CrossRef]

23. Hordoir, R.; Axell, L.; Löptien, U.; Dietze, H.; Kuznetsov, I. Influence of sea level rise on the dynamics of salt inflows in the Baltic Sea. *J. Geophys. Res. Ocean.* **2015**, *120*, 6653–6668. [CrossRef]

24. Wekerle, C.; Wang, Q.; Danilov, S.; Schourup-Kristensen, V.; von Appen, W.J.; Jung, T. Atlantic Water in the Nordic Seas: Locally eddy-permitting ocean simulation in a global setup. *J. Geophys. Res. Ocean.* **2017**, *122*, 914–940. [CrossRef]

25. Androsov, A.; Fofonova, V.; Kuznetsov, I.; Danilov, S.; Rakowsky, N.; Harig, S.; Brix, H.; Helen Wiltshire, K. FESOM-C v.2: Coastal dynamics on hybrid unstructured meshes. *Geosci. Model Dev.* **2019**, *12*, 1009–1028. [CrossRef]

26. Fofonova, V.; Androsov, A.; Sander, L.; Kuznetsov, I.; Amorim, F.; Hass, H.C.; Wiltshire, K.H. Non-linear aspects of the tidal dynamics in the Sylt-Rømø Bight, south-eastern North Sea. *Ocean Sci.* **2019**, *15*, 1761–1782. [CrossRef]

27. Danilov, S.; Androsov, A. Cell-vertex discretization of shallow water equations on mixed unstructured meshes. *Ocean Dyn.* **2015**, *65*, 33–47. [CrossRef]

28. OSPAR Commission London. *Quality Status Report 2010*; Technical report; OSPAR Commission London: London, UK, 2010.

29. Clark, S.; Schroeder, F.; Baschek, B. *The Influence of Large Offshore Wind Farms on the North Sea And Baltic Sea—A Comprehensive Literature Review*; Technical report; Helmholtz-Zentrum Geesthacht, Zentrum für Material-und Küstenforschung: Geesthacht, Germany, 2014.

30. Sündermann, J.; Pohlmann, T. A brief analysis of North Sea physics. *Oceanologia* **2011**, *53*, 663–689. [CrossRef]

31. Reise, K.; Baptist, M.; Burbridge, P.; Dankers, N.; Fischer, L.; Flemming, B.; Oost, A.P.; Smit, C. *The Wadden Sea—A Outstanding Tidal Wetland. Wadden Sea Ecosystem No. 29*; Technical Report 29; Common Wadden Sea Secretariat (CWSS): Wilhelmshaven, Germany, 2010.

32. Stanev, E.V.; Schulz-Stellenfleth, J.; Staneva, J.; Grayek, S.; Grashorn, S.; Behrens, A.; Koch, W.; Pein, J. Ocean forecasting for the German Bight: From regional to coastal scales. *Ocean Sci.* **2016**, *12*, 1105–1136. [CrossRef]

33. Haller, M.; Janssen, F.; Siddorn, J.; Petersen, W.; Dick, S. Evaluation of numerical models by FerryBox and fixed platform in situ data in the southern North Sea. *Ocean Sci.* **2015**, *11*, 879–896. [CrossRef]

34. Purkiani, K.; Becherer, J.; Flöser, G.; Gräwe, U.; Mohrholz, V.; Schuttelaars, H.M.; Burchard, H. Numerical analysis of stratification and destratification processes in a tidally energetic inlet with an ebb tidal delta. *J. Geophys. Res. Ocean.* **2015**, *120*, 225–243. [CrossRef]

35. Stanev, E.V.; Pein, J.; Grashorn, S.; Zhang, Y.; Schrum, C. Dynamics of the Baltic Sea straits via numerical simulation of exchange flows. *Ocean Model.* **2018**, *131*, 40–58. [CrossRef]

36. Zhang, Y.J.; Stanev, E.; Grashorn, S. Unstructured-grid model for the North Sea and Baltic Sea: Validation against observations. *Ocean Model.* **2016**, *97*, 91–108. [CrossRef]

37. Pein, J.U.; Stanev, E.V.; Zhang, Y.J. The tidal asymmetries and residual flows in Ems Estuary. *Ocean Dyn.* **2014**, *64*, 1719–1741. [CrossRef]

38. Shom. Shom (Service Hydrographique et Océanographique de la Marine) (2018). EMODnet Digital Bathymetry (DTM 2018). 2018. Available online: https://sextant.ifremer.fr/record/18ff0d48-b203-4a65-94a9-5fd8b0ec35f6/ (accessed on 12 May 2020).

39. Geuzaine, C.; Remacle, J.F. Gmsh: A 3-D finite element mesh generator with built-in pre- and post-processing facilities. *Int. J. Numer. Methods Eng.* **2009**, *79*, 1309–1331. [CrossRef]

40. Remacle, J.F.; Lambrechts, J.; Seny, B.; Marchandise, E.; Johnen, A.; Geuzainet, C. Blossom-Quad: A non-uniform quadrilateral mesh generator using a minimum-cost perfect-matching algorithm. *Int. J. Numer. Methods Eng.* **2012**, *89*, 1102–1119. [CrossRef]

41. Nú nez-Riboni, I.; Akimova, A. Monthly maps of optimally interpolated in situ hydrography in the North Sea from 1948 to 2013. *J. Mar. Syst.* **2015**, *151*, 15–34. [CrossRef]

42. Egbert, G.D.; Erofeeva, S.Y. Efficient inverse modeling of barotropic ocean tides. *J. Atmos. Ocean. Technol.* **2002**, *19*, 183–204.<0183:EIMOBO>2.0.CO;2. [CrossRef]

43. Androsov, A.A.; Klevanny, K.A.; Salusti, E.S.; Voltzinger, N.E. Open boundary conditions for horizontal 2-D curvilinear-grid long-wave dynamics of a strait. *Adv. Water Resour.* **1995**, *18*, 267–276. [CrossRef]

44. Ridal, M.; Olsson, E.; Unden, P.; Zimmermann, K.; Ohlsson, A. *Deliverable D2.7: HARMONIE Reanalysis Report of Results and Dataset*; Technical report; SMHI: North Köping, Sweden, 2017.

45. Kerimoglu, O.; Hofmeister, R.; Maerz, J.; Riethmüller, R.; Wirtz, K.W. The acclimative biogeochemical model of the southern North Sea. *Biogeosciences* **2017**, *14*, 4499–4531. [CrossRef]

46. Radach, G.; Pätsch, J. Variability of continental riverine freshwater and nutrient inputs into the North Sea for the years 1977-2000 and its consequences for the assessment of eutrophication. *Estuaries Coasts* **2007**, *30*, 66–81. [CrossRef]

47. Voynova, Y.G.; Brix, H.; Petersen, W.; Weigelt-Krenz, S.; Scharfe, M. Extreme flood impact on estuarine and coastal biogeochemistry: The 2013 Elbe flood. *Biogeosciences* **2017**, *14*, 541–557. [CrossRef]

48. Valle-Levinson, A.; Stanev, E.; Badewien, T.H. Tidal and subtidal exchange flows at an inlet of the Wadden Sea. *Estuarine Coast. Shelf Sci.* **2018**, *202*, 270–279. [CrossRef]

49. Stanev, E.V.; Al-Nadhairi, R.; Valle-Levinson, A. The role of density gradients on tidal asymmetries in the German Bight. *Ocean Dyn.* **2015**, *65*, 77–92. [CrossRef]

50. Codiga, D.L. *Unified Tidal Analysis and Prediction Using the UTide Matlab Functions*; Technical report; Graduate School of Oceanography, University of Rhode Island: Narragansett, RI, USA, 2011. [CrossRef]

51. Baschek, B.; Schroeder, F.; Brix, H.; Riethmüller, R.; Badewien, T.H.; Breitbach, G.; Brügge, B.; Colijn, F.; Doerffer, R.; Eschenbach, C.; et al. The Coastal Observing System for Northern and Arctic Seas (COSYNA). *Ocean Sci.* **2017**, *13*, 379–410. [CrossRef]

52. Bersch, M.; Gouretski, V.; Sadikni, R.; Hinrichs, I. KLIWAS North Sea Climatology of Hydrographic Data (Version 1.0). *World Data Cent. Clim. (WDCC)* **2013**.10.1594/$WDCC/KNSC_hyd_v$1.0. [CrossRef]

53. Purkiani, K.; Becherer, J.; Klingbeil, K.; Burchard, H. Wind-induced variability of estuarine circulation in a tidally energetic inlet with curvature. *J. Geophys. Res. Ocean.* **2016**, *121*, 3261–3277. [CrossRef]

54. Becherer, J.; Stacey, M.T.; Umlauf, L.; Burchard, H. Lateral Circulation Generates Flood Tide Stratification and Estuarine Exchange Flow in a Curved Tidal Inlet. *J. Phys. Oceanogr.* **2014**, *45*, 638–656. [CrossRef]

55. Petersen, W.; Schroeder, F.; Bockelmann, F.D. FerryBox—Application of continuous water quality observations along transects in the North Sea. *Ocean Dyn.* **2011**, *61*, 1541–1554. [CrossRef]

56. Petersen, W. FerryBox systems: State-of-the-art in Europe and future development. *J. Mar. Syst.* **2014**, *140*, 4–12. [CrossRef]

57. Grayek, S.; Staneva, J.; Schulz-Stellenfleth, J.; Petersen, W.; Stanev, E.V. Use of FerryBox surface temperature and salinity measurements to improve model based state estimates for the German Bight. *J. Mar. Syst.* **2011**, *88*, 45–59. [CrossRef]

58. Dieterich, C.; Wang, S.; Schimanke, S.; Gröger, M.; Klein, B.; Hordoir, R.; Samuelsson, P.; Liu, Y.; Axell, L.; Höglund, A.; et al. Surface Heat Budget over the North Sea in Climate Change Simulations. *Atmosphere* **2019**, *10*, 272. [CrossRef]

59. Koldunov, N.V.; Aizinger, V.; Rakowsky, N.; Scholz, P.; Sidorenko, D.; Danilov, S.; Jung, T. Scalability and some optimization of the Finite-volumE Sea ice-Ocean Model, Version 2.0 (FESOM2). *Geosci. Model Dev.* **2019**, *12*, 3991–4012. [CrossRef]

 *water*

Article

Emulation of a Process-Based Salinity Generator for the Sacramento–San Joaquin Delta of California via Deep Learning

Minxue He *, Liheng Zhong †, Prabhjot Sandhu and Yu Zhou

California Department of Water Resources, 1416 9th Street, Sacramento, CA 95314, USA;
lihengzhong@berkeley.edu (L.Z.); Prabhjot.Sandhu@water.ca.gov (P.S.); Yu.Zhou@water.ca.gov (Y.Z.)
* Correspondence: kevin.he@water.ca.gov; Tel.: +1-916-651-9634
† Now at Descartes Labs, San Francisco, CA 94103, USA.

Received: 30 June 2020; Accepted: 22 July 2020; Published: 23 July 2020

Abstract: Salinity management is a subject of particular interest in estuarine environments because of the underlying biological significance of salinity and its variations in time and space. The foremost step in such management practices is understanding the spatial and temporal variations of salinity and the principal drivers of these variations. This has traditionally been achieved with the assistance of empirical or process-based models, but these can be computationally expensive for complex environmental systems. Model emulation based on data-driven methods offers a viable alternative to traditional modeling in terms of computational efficiency and improving accuracy by recognizing patterns and processes that are overlooked or underrepresented (or overrepresented) by traditional models. This paper presents a case study of emulating a process-based boundary salinity generator via deep learning for the Sacramento–San Joaquin Delta (Delta), an estuarine environment with significant economic, ecological, and social value on the Pacific coast of northern California, United States. Specifically, the study proposes a range of neural network models: (a) multilayer perceptron, (b) long short-term memory network, and (c) convolutional neural network-based models in estimating the downstream boundary salinity of the Delta on a daily time-step. These neural network models are trained and validated using half of the dataset from water year 1991 to 2002. They are then evaluated for performance in the remaining record period from water year 2003 to 2014 against the process-based boundary salinity generation model across different ranges of salinity in different types of water years. The results indicate that deep learning neural networks provide competitive or superior results compared with the process-based model, particularly when the output of the latter are incorporated as an input to the former. The improvements are generally more noticeable during extreme (i.e., wet, dry, and critical) years rather than in near-normal (i.e., above-normal and below-normal) years and during low and medium ranges of salinity rather than high range salinity. Overall, this study indicates that deep learning approaches have the potential to supplement the current practices in estimating salinity at the downstream boundary and other locations across the Delta, and thus guide real-time operations and long-term planning activities in the Delta.

Keywords: salinity; deep learning; martinez boundary salinity generator; Sacramento–San Joaquin Delta

1. Introduction

Salinity has been long recognized as a critical environmental variable in estuaries which are transition zones between upstream freshwater environments and downstream saline marine environments [1]. The spatial and temporal variation pattern of salinity in estuaries plays a dominant role in the health of estuarine habitats and biota [2–7]. This pattern is typically influenced by drivers including upstream freshwater inflow, downstream tidal forces, as well as local water diversions, precipitation, evaporation, and wind in the estuaries, among others. Understanding the variation pattern of salinity is the foremost step in predicting its future behavior and thus guiding salinity management in estuaries [8–12]. This is particularly the case for estuarine environments with paramount economic, ecological, and social significance including the Sacramento–San Joaquin Delta in California, United States.

As the fifth largest economy in the world, the state of California accommodates a population of nearly 40 million and is one of the most productive agricultural areas globally [13]. Reliable water supply is indispensable to support such a large population and sustain such a robust economy. However, the spatial and temporal distribution of precipitation, the largest water supply source for the state, largely mismatches water demands. Most of the state's population and farmlands (and thus water demand) is in the southern half, while most of the precipitation falls in the northern mountain ranges in the state. In addition, a majority of precipitation occurs in the wet winter season while the highest water demand is normally in the dry summer and fall. To balance the mismatch, a complex water storage and transfer system has been built in the state to redistribute water across different spatial and temporal scales. The most critical components of the system are the State Water Project (SWP) and the Central Valley Project (CVP) which are operated by the state and the federal governments, respectively. SWP and CVP infrastructure consist of tens of dams and reservoirs, pumping plants, hydro-power generation plants, and over 1000 km of aqueducts, tunnels, canals, and pipelines [14,15].

The hub of this statewide water redistribution system is the Sacramento–San Joaquin Delta (Delta, California, CA, USA). Physically, the delta is a patchwork of islands surrounded by about 1100 km of waterways (Figure 1). It receives freshwater from the largest two rivers in the state, namely Sacramento River on the north and San Joaquin River and its tributaries on the south-east. Freshwater inflows are either diverted to water users within and outside of the delta or serve to repel seawater intrusion from its downstream boundary at Martinez (Figure 1). Ecologically, the delta is a globally important biodiversity hot spot with the highest priority of conservation [16]. It provides habitats that support about 750 species of plants and animals including some near extinction [17]. Socioeconomically, the delta provides water to about two thirds of the state's population and over 15,000 km^2 of farmlands via SWP and CVP deliveries. Millions of people use the delta for recreation and transportation [18]. These physical, ecological, and socioeconomic features of the delta drive SWP and CVP operations with the coequal goals of a reliable water supply and an ecologically sustainable delta ecosystem [19]. The SWP and CVP pump water from southern delta (Figure 1) and transfer the water to municipal and agricultural users in the state. The pumping time and rates are dictated by state and federal regulatory requirements to ensure that: (1) flow and water quality standards at various locations in the delta are complied with, and that (2) additional regulations to protect endangered species are followed [20,21]. One critical water quality standard is that the salinity level, typically reported in units of electrical conductivity (EC) as microSiemens/cm (μs/cm), at compliance locations cannot exceed preset threshold values during certain periods in a certain type (e.g., wet, dry, critical) of water year. Traditionally empirical and process-based models have been used to simulate salinity variations at these locations to guide real-time delta operations (e.g., SWP and CVP operations) and long-term planning studies (e.g., structural changes to the delta) to ensure salinity compliance.

One of the earliest models developed for this purpose is the conceptual–empirical salinity gradient model (i.e., G-model) of [22]. The model derives salinity from antecedent delta outflow based on the assumption that there is a non-linear relationship between these variables. Hutton et al. [11] extended the G-model to simulate the low salinity zone in the delta defined as the position of a predetermined salinity isohaline. Numerical process-based models have also been developed to simulate the spatial and temporal variation of salinity in the delta. These models include, to name a few, but are not limited to, the one-dimensional Delta Simulation Model II (DSM2) [23], two-dimensional RMA10 [24] and TRIM2D [25], and three-dimensional SCHISM [26,27], UnTrim [28,29], and SUNTANS [30,31]. Although physically more rigorous and being able to provide higher spatial resolution on flow and salinity distribution than one-dimensional models, multi-dimensional models are computationally expensive. For studies with long data temporal sequences and multiple scenarios involved, simpler one-dimensional models are still favored. This is the case for DSM2 which is widely applied in contemporary applications in the delta [32]. The DSM2 domain covers the entire delta with Martinez as its downstream boundary. For operational planning and forecasting studies, DSM2 relies on the Martinez Boundary Salinity Generator (MBSG) [33] to produce downstream salinity boundary conditions for DSM2 simulations.

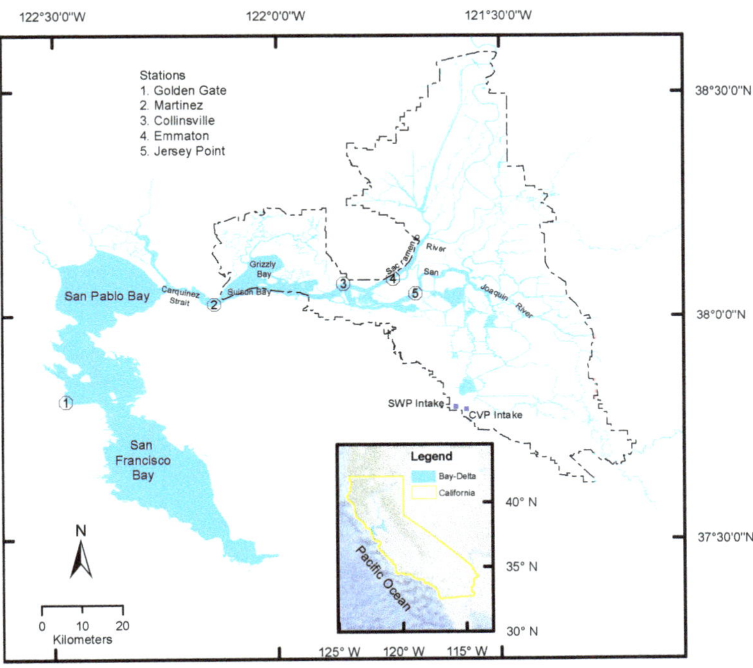

Figure 1. Location map showing the San Francisco Bay and Sacramento–San Joaquin Delta (Bay-Delta) Estuary and the study location Martinez along with several other key water quality stations including the Golden Gate, Collinsville, Emmaton, and Jersey Point. The dash line highlights the modeling domain of the Delta Simulation Model II (DSM2). The insert map illustrates the location of the Bay-Delta in California.

In addition to empirical and process-based models, data-driven models including Artificial Neural Networks (ANNs) have also been explored in deriving salinity in the delta area [34–39]. An ANN employs a mathematical network structure to implicitly identify the relationships between one or more inputs (e.g., usually measured variables such as stage or flow) and outputs (e.g., salinity at selected locations) datasets. The basic processing units in the network are called neurons. These neurons are arranged in layers and are connected to other neurons in adjacent layers. Multilayer perceptrons (MLPs) are probably the most popular ANN models applied in the field of water resources engineering [40]. The above-mentioned ANN studies generally use MLP-based models in estimating salinity in the Delta. An MLP is a feedforward ANN typically consisting of two visible layers on both ends of the network (i.e., input and output layers) and one or more hidden layers in the middle. A neuron in a specific layer takes inputs from neurons in the previous layer and outputs a linear or non-linear transformation of the combined input information to neurons in the next layer. The connections between neurons are represented by linear weights. These weights are determined in the training process by minimizing the difference (i.e., error signal) between network predictions of the variable of interest (e.g., salinity) and the corresponding observations. The most common training method is gradient descent, which propagates the difference backward into the network and updates the weights according to the chain rule [41].

Despite their popularity, MLPs do not treat the sequential ordering of input time series as a feature during training. For non-linear systems where short- or long-term temporal dependencies exist between output and input (e.g., salinity at the current time relates to antecedent flows), MLPs may not be the most viable choice [42]. Recurrent Neural Networks (RNNs), a different category of ANN, are designed to overcome this drawback of MLPs [43]. RNNs process input in its temporal order. The output of hidden layer neurons at each time step is recurrently fed as an additional input to the next time step. This feature grants RNNs the advantage of better understanding the temporal dynamics between input and output variables. In spite of this advantage, standard RNN models are shown to have difficulties in capturing long-term dependencies [44]. This is mainly caused by two problems encountered during the training process: vanishing gradient (network weights approach zero) and exploding gradient (network weights become extremely large). These two problems occur mostly because the error signal can only be back propagated effectively for a few steps [45].

Many variants of RNNs have been proposed to avoid the vanishing and exploding gradient problems. The most well-known and successful variant is the long short-term memory (LSTM) network [45]. LSTM introduces the concept of gates, which are essentially neurons with learnable weights. The gates inform the network on what information to discard, what to retain, and for how long. This gate configuration helps the network preserve essential information over a long time and avoid rapid error signal decay. LSTM networks have only been applied recently in the field of water resources in terms of modeling rainfall-runoff process [46–49], groundwater table [50,51], water level in channels [52], water quality [53,54], and reservoir operations [55]. Given the long-term dependencies between salinity and flow/stage, LSTM should also be suitable for salinity simulation given flow and stage inputs.

Convolutional neural network (CNN) is another category of ANNs. A CNN consists of a sequence of layers that shrink in length from the input layer to the output layer. The shrinking aims to condense information learnt from previous layers to more abstract concepts in deep layers [42]. CNN is a leading network architecture in deep learning techniques and has had extremely successful applications in image pattern recognition and classification [56–59]. It has only recently received significant attention in water-related time series modeling in terms of ground water level prediction [60], precipitation estimation [61,62], and flood forecasting [63,64].

To our knowledge, few studies have explored the applicability of state-of-the-art deep learning techniques (e.g., LSTM, CNN) in salinity estimation for deltaic and/or estuarine environments, not to mention in the Sacramento–San Joaquin Delta specifically. This study explores the capacity of these techniques by presenting a case study of emulating the Martinez Boundary Salinity Generator in estimating the downstream salinity boundary (i.e., Martinez salinity) for the delta via LSTM- and CNN-based models. The salinity generator itself is applied as the benchmark model. A number of MLP-based neural network models are also proposed for comparison purpose. The rest of the paper is organized as follows: Section 2 describes the study area, the available dataset, the Martinez Salinity Boundary Generator, the proposed neural network models, and the evaluation metrics; Section 3 presents the results and findings; Section 4 discusses data stationarity, study limitations, implications, and future work; the last section concludes the paper.

2. Materials and Methods

2.1. Study Location and Dataset

The Sacramento–San Joaquin Delta (Delta) is the hub of California's vast water supply system with critical urban, agricultural, environmental, industrial, and recreational importance. The Delta is an estuary at the confluence of the largest two river systems in California, the rain-dominated Sacramento River in the north and snow-dominated San Joaquin River in the south (Figure 1). These two rivers and their tributaries contribute freshwater to the Delta. Freshwater is the main source for Delta diversions (including SWP and CVP exports) and Delta consumptive use. It is also used to repel the intrusion of seawater which enters the San Francisco Bay and San Pablo Bay via the Golden Gate (Station 1 in Figure 1). The downstream boundary of the Delta, Martinez (Station 2 in Figure 1), is connected to the San Pablo Bay via the Carquinez Strait. Martinez is under strong influence of salty tides from the San Pablo Bay. Salinity at Martinez serves as the salinity boundary for the Delta.

SWP and CVP pump water from south Delta for export to serve over 25 million people (about two thirds of the state's population) and 15,000 km^2 farmland in California. Water quality standards have been developed to ensure that the water at the intakes of SWP and CVP is appropriate for drinking water, agricultural, and other purposes [20,21]. In California, five types of water years are defined to facilitate water resources management. They are wet (W), above-normal (AN), below-normal (BN), dry (D), and critical (C) years and are defined according to the overall wetness of a specific year [20]. The water quality standards vary across different water year types. Salinity is being monitored at a range of key compliance stations including Collinsville, Emmaton, and Jersery Point (Stations 3, 4, and 5 in Figure 1) to ensure compliance with these water quality standards. For instance, in below-normal years, the salinity (represented by electrical conductivity (EC)) at Jersey Point should remain below 450 µs/cm from 1 April to 20 June and below 740 µs/cm from 21 June to 15 August. Table A1 in the Appendix A provides a more detailed list of such standards at Jersey Point and Emmaton. Martinez salinity is the major salinity source for these compliance stations. To have a clear understanding on the salinity at these locations and thus the overall compliance status in the Delta, it is imperative to have a rigorous estimate on Martinez salinity. This is particularly true in planning studies (e.g., different operation options or different structural change scenarios in the Delta) where no field measurements of salinity at the interior locations would be available.

This study utilizes the same dataset as applied in the [65] study. The dataset includes a 24 year period (water year 1991–2014) of daily observed water stage (average, minimum, and maximum) at Martinez, Martinez salinity (hereinafter "reference salinity") and the net Delta outflow (NDO) calculated based on observed or modeled inflows and water uses in the Delta [66]. The salinity exhibits a strong seasonality with the lowest value in early spring (Figure 2). It increases gradually till peaking typically near the end of fall. In the winter, when upstream reservoirs in the Sacramento and San Joaquin River system increase releases to reserve storage to manage wet season floods, the salinity at Martinez starts dropping. The NDO shows a roughly reversed pattern with its peak in the winter and its minimum in the fall. It is clear that variations in salinity mimic that of the NDO, although in reverse fashion as expected. There is a negative correlation (with a Pearson correlation coefficient, R = −0.91) between them on the annual scale. The average stage has a different cyclic pattern throughout the year, owing to a more complex relationship with both incoming freshwater as well as actual cycles in tidal elevation. Its correlation with the salinity is much weaker (R = 0.11). The variation patterns of these three variables are also evident when looking at their daily time series during the entire 24 year period (Figure A1 in the Appendix A). In this study, the first 12 year period (water year 1991–2002) is used as the training/validation period for neural network models, while the second 12 year period (water year 2003–2014) is used as the evaluation period.

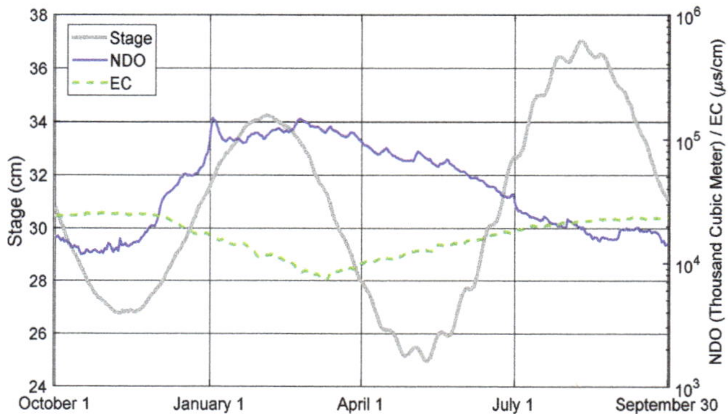

Figure 2. Long-term (water year 1991–2014) average daily Martinez water stage (Stage; left *y*-axis), net Delta outflow (NDO; right *y*-axis) and Martinez salinity represented by Electrical Conductivity (EC; right *y*-axis).

2.2. Martinez Boundary Salinty Generator

The Martinez Boundary Salinity Generator (MBSG) integrates the conceptual–empirical G-model of [22] and a linear filter proposed by [33] for planning and forecasting studies involved with DSM2 modeling. The G-model simulates antecedent flow-salinity relationship as follows:

$$S_t = (S_O - S_U) \times e^{-\alpha G_t^k} + S_U \tag{1}$$

where S_t is the salinity at time step t; S_O and S_U represent the downstream ocean and upstream river salinity, respectively; α and k are a dispersion parameter (with the effect of upstream distance consolidated) and an empirical shape parameter, respectively; and G is a function representing the antecedent flow defined as:

$$\frac{dG}{dt} = \frac{(Q - G) \times G}{\beta} \tag{2}$$

where β is an empirical constant and Q is the volumetric flow rate which is net Delta outflow for Martinez.

The linear filter models tidally varying effects from the ocean based on the assumption that "tidally-averaged salinity is the result of a uniform, harmonic advection acting on the exponential salinity profile from G-model" [33]. This study provides the mathematical forms of the filter as well as the updated salinity estimation equation for simplicity. For detailed explanation on the theory, implementation, and application of the linear filter in estimating Martinez salinity, the readers are referred to [33]. Specifically, the change made to the G-model is that a harmonic position reflecting displacement of the salinity profile, x_t, is added to Equation (1):

$$S_t = (S_O - S_U) \times e^{-\widetilde{\alpha} G_t^k x_t} + S_U \tag{3}$$

where $\widetilde{\alpha}$ is the decay parameter in Equation (1) before distance is bundled into it. x_t can be divided into a centered position (x_0) and a harmonic perturbation (x_t'): $= x_0 + x_t'$. Combining $-\widetilde{\alpha}x_0$ into a coefficient β_1 and rearranging Equation (3) yield:

$$ln(\frac{S_t - S_U}{S_O - S_U}) = \beta_1 G_t^k + x_t' G_t^k \tag{4}$$

x'_t can be written as a convolution filter modeling displacement on lagged Martinez stage:

$$x'_t = \sum_{i=0}^{n_i} a_i z_{t+i_0\Delta t - i\Delta t} \tag{5}$$

where z_t is the tide stage; a_i are the filter coefficients; n_i represents the length of the convolution kernel (i.e., number of lagged input values applied); and Δt stands for the spacing between lagged values; i_0 designates an offset of the filter. With Equation (5) incorporated, the governing equation of MBSG becomes:

$$ln(\frac{S_t - S_U}{S_O - S_U}) = \beta_1 G_t^k + G_t^k \sum_{i=0}^{n_i} a_i z_{t+i_0\Delta t - i\Delta t} \tag{6}$$

The MBSG was recently recalibrated [65] using an automated parameter optimization software named Parameter Estimation (PEST) [67]. The recalibration improves model performance when compared to the original calibration [33], particularly in the high salinity range. This study uses the PEST calibrated MBSG as the baseline model to benchmark the proposed neural network models.

2.3. Artificial Nerual Networks

Multi-Layer Perceptron (MLP) is the plain form of neural networks. In MLP, each neuron in each layer is fully connected to all neurons in adjacent layers. An MLP model with one or more hidden layers is often used to evaluate the baseline performance of deep neural networks without employing any special architecture. Specialized network architectures have been developed, and the most popular ones include Recurrent Neural Network (RNN), which is naturally suitable to handle sequential data, and Convolutional Neural Network (CNN), which captures patterns in a hierarchical manner. A widely used special form of RNN is Long Short-Term Memory (LSTM), in which neurons are organized as sequential units each of which "remembers" values over arbitrary time steps, long or short. One-dimensional convolutional neural network (Conv1D) is a special form of CNN. Conv1D employs layers of one-dimensional filters to capture hierarchical patterns in series data including time series. By stacking many convolutional layers, Conv1D can effectively combine local and overall patterns to learn complex temporal features which are very hard to delineate with pre-defined mathematical models.

In this study, the output of the prediction task of neural network models is the EC of the current day. For each model used in this task, different sets of inputs were tested. Candidate input variables are daily NDO and mean, minimum, and maximum stage in the previous 60 days (including the current day) as well as the simulated EC by the PEST–calibrated MBSG model at the current day. The selection of 60 days is empirical. In the Delta, salinity is influenced by antecedent flow in the preceding several months. However, after about two months, the influence generally becomes very weak (with an absolute correlation value less than 0.5; Figure A2). We tested shorter and longer periods. The results were not as ideal as that of the case when 60 day is applied. Daily NDO is a basic indicator of the hydrologic condition in Delta, and the statistics of stage level observation provide more detailed information on daily stage dynamics. Three scenarios with different combinations of input variables were investigated. In Scenario 1, the input variables include daily NDO and average stage. Scenario 2 also employs the two daily series and adds daily minimum and maximum stage to further delineate daily stage dynamics. In Scenario 3, the input variables include all inputs in Scenario 2 as well as the simulated EC by the MBSG model.

MLP and LSTM are applied to the three scenarios of input variables. For MLP, the input size is 120 (2 × 60, Scenario 1), 240 (4 × 60, Scenario 2), or 241 (4 × 60 + 1, Scenario 3). For LSTM, the main data input is expected to be time series with equal length for which the simulated EC at the current day does not fit. The input size to the neural network is 120 (2 × 60) for Scenario 1 and 240 (4 × 60) for Scenarios 2 and 3. As the simulated EC from the MBSG model is available in Scenario 3, the neural network predicts the relative difference between the simulated EC and the actual EC, using the simulated EC by the

MBSG model as an additional input to the last layer of the network. The neural network models used in Scenarios 1 and 2 predict the absolute level of EC. The special use of the simulated EC is because of the requirement of LSTM to have aligned series as inputs. In addition to looking at impacts of different input variables on network model performance, the impacts of different network hyper-parameters are also investigated. Hyper-parameters of MLP and LSTM configured with Scenario 3 input variables are fined-tuned to yield a fourth MLP and a fourth LSTM model, respectively. Table 1 lists all models investigated in this study.

Table 1. Study Models.

Model Name	Description
MBSG	PEST calibrated MBSG model in the [65] study
MLP1	MLP with NDO and average stage as input
MLP2	MLP with MLP1 inputs plus maximum and minimum stage
MLP3	MLP with MLP2 inputs plus MBSG simulated Martinez salinity
MLP4	MLP with MLP3 inputs and fine-tuned network parameters
LSTM1	LSTM with NDO and average stage as input
LSTM2	LSTM with LSTM 1 inputs plus maximum and minimum stage
LSTM3	LSTM with LSTM 2 inputs plus MBSG simulated Martinez salinity
LSTM4	LSTM with LSTM inputs and fine-tuned network parameters
Hybrid	Hybrid MBSG and Conv1D

We also test the feasibility of combining the existing process-based model and deep learning architectures as a hybrid model. One of the key steps in the MBSG model is to reduce the recent stage level series to a scalar as an indicator of current hydrodynamic conditions. The existing approach in the MBSG model is using 15 min stage data in the past 72 h at a 12 h interval, or 7 stage observations (i.e., n_i in Equation (5)) in total to quantify the relationship between short-term stage dynamics and salinity. Traditionally it is very hard to fully utilize the temporal information in the dense and noisy 15 min series. Although human experts may interpret the 15 min series to some extent, building explicit rules for model development is prohibitive. As a result, the existing approach in the MBSG model only samples the 15 min every 12 h to simplify the calculation. Conv1D is particularly capable of learning very complex patterns from one-dimensional data and the learning process requires minimal human input. We replaced the existing 12 h sampling approach with a stack of Conv1D layers which take the original 15 min series as inputs, hoping to discover and employ the patterns in the denser series data that might be neglected in the past. In this way we have a hybrid model, which includes most of the physical processes of the MBSG model as well as a deep learning-based pattern recognizer to deal with the complexity in dense stage observations.

Data from water year 1991 to 1999 are utilized to train the neural network models. To optimize hyper-parameters and select the optimal network structure of a certain type, data in water years 2000, 2001, and 2002 are used as a validation set which is not directly used in training but in the evaluation of performance by various combinations of hyper-parameters. Hyper-parameters of MLP include the number of hidden layers and the number of neurons in each layer. For LSTM, hyper-parameters include the number of LSTM units/layers, the number of filters in the recurrent units, the dropout rate between layers, and the dropout rate between time steps. These parameters are specified in Table A2 in the Appendix A. For the Conv1D component in the hybrid model, we tuned the number of Conv1D layers, the number of filters in each layer, and the dropout rate. We also tuned batch size and initial learning rate for all types of deep networks. The Adam optimizer was used in all experiments [68].

2.4. Study Metrics

This study employs five metrics which provide complementary information on the performance of the proposed models in simulating Martinez salinity. They are defined in Table 2 where S represents salinity, $\overline{S}$ means average salinity, t indicates a specific time step, n stands for the total number of time steps, and *sim* and *ref* designate simulated and reference values, respectively. Specifically, these metrics include the commonly used percent bias and mean absolute error between the reference salinity and the corresponding model simulations. Percent bias shows whether the model over-estimates or under-estimates the reference salinity by how much on average sense. Mean absolute error indicates the average magnitude of model simulation errors. In addition, the study also includes three metrics that represent the three components of the Taylor Diagram: standard deviation (SD), correlation coefficient (R), and centered root mean square difference (RMSD). The Taylor Diagram provides a concise summary of how well different model simulations match the reference field in terms of these three components in a single diagram [69]. The correlation coefficient measures the strength of the linear relationship between model simulations and the reference. The standard deviation quantifies the amplitude of their variations. The centered root mean square difference provides the centered (with mean subtracted) model error.

Table 2. Study Metrics.

Name	Description	Formula
Bias	Percent bias	$\dfrac{\sum_{t=1}^{n}\left(S_{sim,t}-S_{ref,t}\right)}{\sum_{t=1}^{n}S_{ref,t}}\times 100$
MAE	Mean Absolute Error	$\dfrac{1}{n}\sum_{t=1}^{n}\left|S_{sim,t}-S_{ref,t}\right|$
SD	Standard Deviation	$\sqrt{\dfrac{\sum_{t=1}^{n}\left(S_{t}-\overline{S}\right)}{n-1}}$
R	Correlation Coefficient	$\dfrac{\sum_{t=1}^{n}\left[\left(S_{sim,t}-\overline{S_{sim}}\right)\times\left(S_{ref,t}-\overline{S_{ref}}\right)\right]}{n\times SD_{sim}\times SD_{ref}}$
RMSD	Centered Root Mean Square Difference	$\sqrt{\dfrac{\sum_{t=1}^{n}\left[\left(S_{sim,t}-\overline{S_{sim}}\right)-\left(S_{ref,t}-\overline{S_{ref}}\right)\right]^{2}}{n}}$

Model results are evaluated in three different ways. Firstly, those five metrics (Table 2) are calculated between model simulated salinity at Martinez and the reference salinity during the entire evaluation period (watery year 2003–2014) to shed light on the overall performance of these models. Secondly, model simulations within different salinity ranges (high, medium, and low) during the entire evaluation period are assessed against the corresponding reference salinity in terms of these five metrics. In practice, there are different management strategies for different salinity ranges. In general, managing high salinity (versus low–medium salinity) is more challenging. Finally, those five metrics are calculated in different water year types to illustrate whether neural network model performance varies with different categories of water years.

3. Results

The results are grouped into three sub-sections accordingly. The first sub-section presents the overall results during the entire evaluation period from water year 2003–2014. In the second sub-section, the entire evaluation period is divided into three sub-periods containing three different ranges (high, medium, and low) of salinity, respectively. Model performance in simulating different ranges of salinity is examined. In the last sub-section, the evaluation period is divided into five sub-periods representing five different water year types, respectively. Model results are evaluated in each of these five sub-periods.

3.1. Entire Evaluation Period

Standard deviation (SD) of simulated salinity at Martinez along with its correlation (R) with the reference salinity as well as its centered root mean square difference (RMSD) for each model are calculated and illustrated in Figure 3. The hybrid MBSG–CNN model slightly outperforms the process-based MBSG model (Figure 3a). The former has a smaller (by an amount of 5.7%) RMSD and a higher (by about 0.4%) R value compared to the latter. The SD values of both models are fairly close to each other. They are both smaller than their counterpart of the reference salinity, indicating that salinity simulations of both models have relatively less variation compared to the reference salinity.

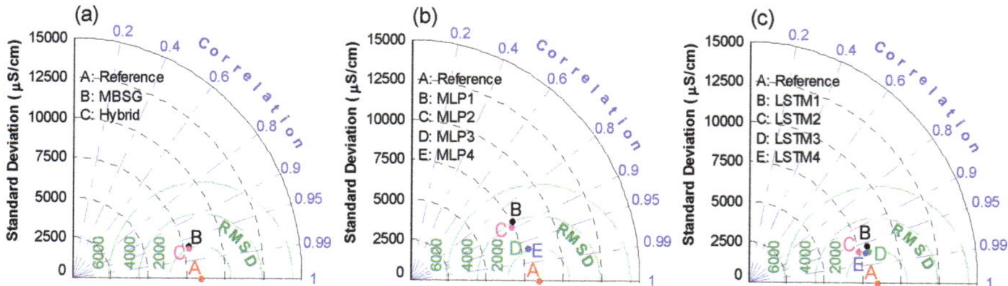

Figure 3. Taylor Diagrams illustrating the correlation (R; the azimuth position), standard deviation (SD; radial distance from the origin, shown on both vertical and horizontal axes), and centered root mean square difference (RMSD; radial distance from the reference point A which serves as the origin for RMSD) between the reference salinity at Martinez and the corresponding salinity simulations generated via (**a**) MBSG and hybrid MBSG–CNN models, (**b**) four MLP models, and (**c**) four LSTM models. Reference point A designates the statistics of the reference salinity itself (RMSD = 0; R = 1). The *X*-axis shows the values of both SD and RMSE of which the ticks are different.

For MLP models, when only using net Delta outflow (NDO) and average water stage as input (MLP1; point B in Figure 3b), the resulting salinity simulations have a smaller (by 9.5%) correlation value and a remarkably larger (by 90%) RMSD compared to MBSG simulations (point B in Figure 3a). The SD values of both MLP1 and MGSB are comparable to each other, yet both are smaller than that of the reference salinity. Adding daily maximum and minimum stage as input (MLP2; point C in Figure 3b) yields simulations with only a slightly higher R value and a marginally smaller RMSD than that of the MLP1 simulations. The SD of MLP2 differs noticeably (12% smaller) from that of the reference salinity. When further incorporating MBSG simulations as an additional input feature (MLP3; point D in Figure 3b), however, the results are improved markedly. The metrics (SD, R, and RMSD) of MLP3 become comparable that of MBSG. Fine-tuning MLP3 hyper-parameters (MLP4; point E in Figure 3b) leads to salinity simulations with even more satisfactory metrics compared to both MLP3 and MBSG.

Different from MLP1, the LSTM model using NDO and average stage information alone as input (LSTM1; point B in Figure 3c) yields comparable simulations to that of the MBSG (point B in Figure 3a). The MBSG model has slightly smaller RMSD and higher R. However, the SD value of LSTM1 is closer to that of the reference salinity compared to the SD value of MBSG. Adding daily maximum and minimum stage information as input (LSTM2; point C in Figure 3c) yields simulations with a higher R value and a lower RMSD than LSTM1. Further including MBSG simulations as input (LSTM3; point D in Figure 3c) leads to salinity simulations with smaller RMSD, higher R, and better SD (i.e., closer to the reference SD) than LSTM2 and MBSG simulations. Fine-tuning LSTM3 hyper-parameters (LSTM4; point E in Figure 3c) results in simulations with even better R and RMSD than that of LSTM3 simulations.

In addition to R, SD, and RMSD, bias, and mean absolute error (MAE) are also calculated for all models studied. Overall, the process-based MBSG model under-estimates the reference salinity (bias = −2.4%; Figure 4). Similarly, most neural network models also under-estimate the salinity except for MLP1 (14.7% bias) and LSTM3 (2.1% bias). In terms of the magnitude, MLP4 and LSTM4 are less biased than MBSG. The remaining neural network models have comparable but slightly higher bias than MBSG except for MLP1. MLP1 also has the largest mean absolute error (MAE = 3979 µs/cm). MLP2 has the second largest MAE value. The MAE values of other models are consistently smaller than 2000 µs/cm. Compared to MBSG, four neural network models, including MLP4, LSTM3, LMST4, and the hybrid model, have smaller MAE values.

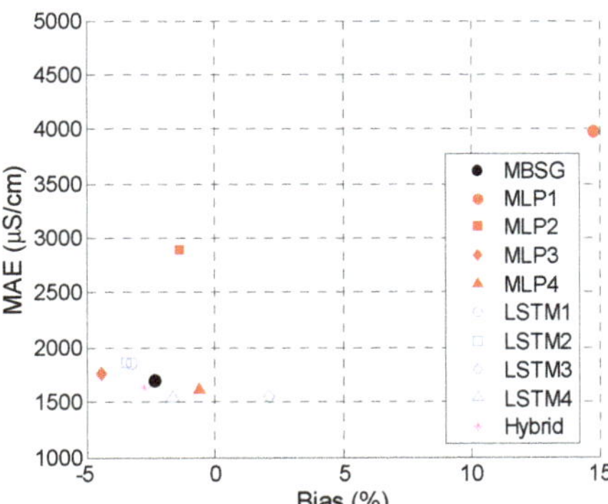

Figure 4. Percent bias (Bias; horizontal axis) and mean absolute error (MAE; vertical axis) between reference and simulated salinity at Martinez via different models during the entire evaluation period.

Looking at five metrics all together, for MLP and LSTM models, incorporating maximum and minimum stage information generally improves network performance. Adding MBSG simulations as an additional network input feature leads to further improvement. The improvement is much more significant for MLP rather than LSTM. Fine-tuning network hyper-parameters is shown to improve the general performance of both MLP and LSTM models. Put differently, among all MLP (LSTM) models, MLP4 (LSTM4) has the best performance in general during the entire evaluation period. Among all nine neural network models, LSTM4 has the smallest RMSD, highest R, and lowest MAE; MLP4 has the lowest bias; LSTM1 and LSTM3 have the closest SD to that of the reference salinity. MLP4, LSTM3, and LSTM4 are the only three models which outperform the process-based MBSG model in terms of all five metrics. In comparison, the hybrid model yields improvement over MBSG in terms of R, RMSD, and MAE. The bias and SD values of the hybrid model are comparable to that of MBSG.

3.2. Different Salinity Ranges

Martinez salinity varies seasonally, typically with low values in winter/spring and high values during summer/fall (Figure 2). Salinity management practices vary accordingly, based on the range of salinity. In addition to looking at model performance in the entire evaluation period, this section further examines its performance during different salinity ranges. Specifically, three ranges are considered, including low range (less than 25th percentile of observed Martinez salinity during the evaluation period; $<1.19 \times 10^4$ microsiemens per centimeter (μs/cm)), medium range (25th percentile to 75th percentile), and high range (over 75th percentile; $>2.53 \times 10^4$ μs/cm). The entire evaluation period is divided into three sub-periods accordingly. The length of the low salinity period is identical to that of the high salinity period, with each accounting for half of the length of the medium range salinity period.

Based on the results during the entire evaluation period presented in Section 3.1, MLP4 and LSTM4 have the best performance among all MLP and LSTM models, respectively. The hybrid model provides generally comparable or superior simulations than MBSG. The current section first evaluates the performance of these three neural network models (MLP4, LSTM4, and Hybrid) against that of the MBSG model (Figure 5). For low range salinity (Figure 5a), all three models yield higher correlation values and lower RMSD compared to MBSG. For medium range salinity (Figure 5b), all three models have higher correlation values and smaller RMSD with one exception. The RMSD of MLP4 is slightly (2%) larger than that of MBSG. For high range salinity (Figure 5c), the RMSD of MLP4 is even higher (by 9.7%) compared to its counterpart of MBSG. The correlation value of MLP4 is also smaller. Conversely, LSTM4 and the hybrid model outperform MBSG in terms of both R and RMSD. Regarding SD, for both medium and high ranges of salinity, all three neural network models and the MBSG model yield simulations with higher variations (higher SD) than the reference salinity; for low salinity, LSTM4 is the only model with a higher than reference SD value. For low, medium, and high ranges of salinity, MLP4, LSTM4, and the hybrid model have the most satisfactory SD (closest to reference SD) values, respectively.

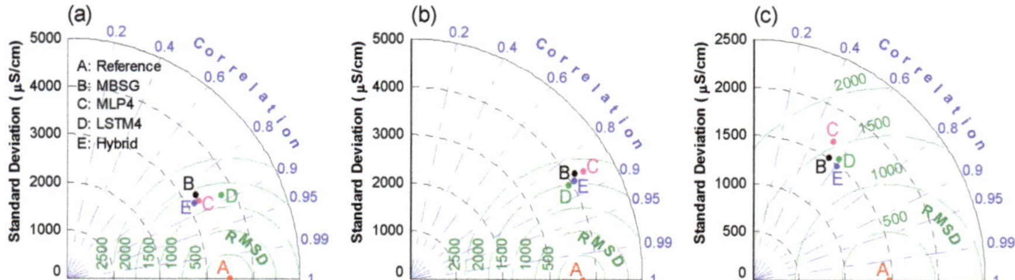

Figure 5. Taylor Diagrams displaying statistics (correlation, standard deviation, and centered root mean square difference) between the reference salinity at Martinez and the corresponding salinity simulations generated via four different models (MBSG, MLP4, LSTM4, and Hybrid MBSG–CNN) grouped in three salinity ranges including (**a**) low salinity range (less than 25% non-exceedance probability), (**b**) medium salinity range (between 25% and 75% non-exceedance probability), and (**c**) high salinity range (above 75% non-exceedance probability).

For the models not depicted in Figure 5, those three metrics (R, SD, and RMSD) are also examined (Table 3). Similar to the results presented in Section 3.1, adding additional information as network input features generally improves model performance across all three salinity ranges. Nevertheless, a noticeable difference is that fine-tuning network hyper-parameters does not necessarily lead to improved performance. The differences in these three metrics between MLP3 (LSTM3) and MLP4 (LSTM4) are minimal. MLP3 performs relatively better than MLP4 in high salinity ranges while LSTM3 generally outperforms LSTM4 in medium and high salinity ranges. Table 3 also indicates that model performance differs evidently in high salinity range versus low to -medium ranges. Specifically, SD, and RMSD values of high salinity simulations are considerably smaller than that of the low and medium salinity simulations while the R value of the former is remarkably smaller than that of the latter. This suggests that simulations on high salinity are generally less spread out (smaller SD and RMSD). However, their linear relationship with the corresponding reference salinity is remarkably weaker when compared to that of simulations on low- to medium ranges of salinity.

Table 3. Statistics of model-simulated Martinez salinity during three different salinity ranges.

Scenarios	Standard Deviation (µs/cm)			Centered Root Mean Square Difference (µs/cm)			Correlation Coefficient		
	Low	Medium	High	Low	Medium	High	Low	Medium	High
Reference	3482	3702	1467	0	0	0	1.000	1.000	1.000
MBSG	3249	4152	1509	1902	2214	1428	0.843	0.847	0.539
MLP1	4428	5392	3306	3504	3837	3306	0.631	0.703	0.222
MLP2	4210	5077	2852	3372	3488	2878	0.630	0.727	0.239
MLP3	3186	4348	1673	1774	2272	1563	0.862	0.853	0.511
MLP4	3258	4346	1681	**1758**	2259	1568	0.866	0.854	0.511
LSTM1	3915	4253	1689	2217	2558	1727	0.827	0.802	0.408
LSTM2	**3648**	3895	1439	1810	2236	1514	0.872	0.828	0.458
LSTM3	4108	**3864**	**1475**	1987	**1992**	**1318**	0.876	0.862	0.598
LSTM4	3734	3920	1565	1773	2005	1376	**0.882**	**0.863**	0.589
Hybrid	3152	4082	1493	1765	2077	1326	0.863	0.862	**0.599**

Bold numbers represent the best metrics.

In terms of bias and MAE, different models perform differently across different salinity ranges. First, all models tend to over simulate low salinity (Figure 6a). The process-based model has a bias of 7.9% and MAE of 1519 µs/cm for low salinity simulations. In comparison, only LSTM1, MLP3, and the hybrid model are less biased among all nine neural network models. Overall, MLP1 is the outlier model with significantly large bias and MAE. MLP2 shows improvement over MLP1. However, its bias and MAE values are still remarkably larger than that of the remaining models. In contrast, MLP3 and the hybrid model have the smallest bias and MAE. Second, all models except for MLP1 under-simulate high salinity (Figure 6c). The bias and MAE of MBSG are −6.4% and 1858 µs/cm, respectively, for high salinity simulations. Four neural network models including MLP4, LSTM1, LSTM3, and LSTM4 have smaller bias and MAE than MBSG. Among them, LSTM3 has the most satisfactory bias and MAE. Finally, most models also under-estimate the medium range salinity (Figure 6b). MLP1 is again the outlier model with the largest positive bias and MAE. LSTM3 is the other model with positive bias (4.2%). MBSG simulations on medium salinity have a bias of −1.2% and MAE of 1723 µs/cm, respectively. In comparison, both MLP4 and LSTM4 have smaller bias and MAE values.

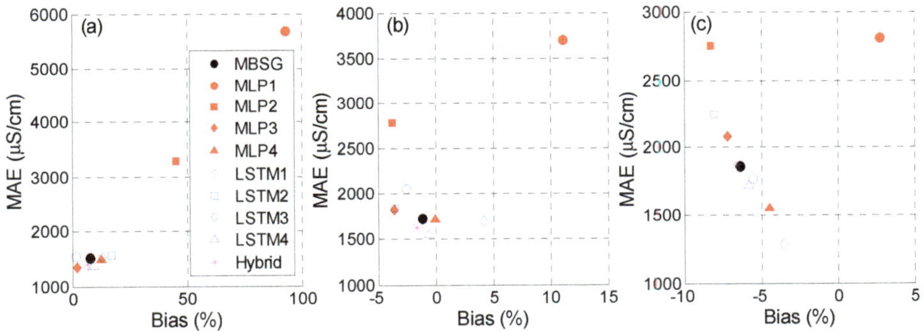

Figure 6. Percent bias (Bias; horizontal axis) and mean absolute error (MAE; vertical axis) between reference and simulated salinity at Martinez via different models grouped in three salinity ranges including (**a**) low salinity range; (**b**) medium salinity range; and (**c**) high salinity range.

All in all, no single model consistently outperforms the others in terms of all five metrics across low, medium, and high ranges of salinity. However, among all models, MLP1 and MLP2 have the worst performance measured by nearly all metrics. For high range salinity, LSTM3 has the best performance in general. It is the least bias model with the smallest RMSD and MAE and the best SD. The associated correlation coefficient (0.598) is very close to the optimal value (0.599) of the hybrid model. For medium range salinity, LSTM3 has the smallest RMSE and the best SD; LSTM4 has the highest correlation coefficient and smallest MAE, while MLP4 is the least biased. The results on low salinity are mixed. The five optimal metrics come from five different models, respectively. Nevertheless, on average, MLP4, LSTM4, and the hybrid models have relatively better performance.

3.3. Different Water Year Types

In the Delta, water quality standards vary with water year types (e.g., Table A1 in the Appendix A). Understanding model performance in different types of water years is critical to guide corresponding salinity management practices. The entire evaluation period (2003–2014) is divided into five sub-periods, with each sub-period containing the data from a specific water year type (W, AN, BN, D, C). There are two wet years, two above-normal years, three below-normal years, three dry years, and two critical years. Therefore, these five sub-periods vary (from two to three years) in length.

Following Section 3.2, this section first examines three metrics illustrated by the Taylor diagram of the process-based MBSG model and three neural network models (MLP4, LSTM4, and Hybrid).

Overall, the performance of these four models are fairly close to each other across all five types of water years (Figure 7). However, none of them consistently outperform the others. Specifically, across all types of water years, LSTM4 and the hybrid model have higher correlation values than MLP4 and MBSG. In addition, the hybrid model has smaller RMSE than MLP4 and MBSG. Regarding SD, MLP4 has the best performance in all types of water years except for above-normal years. The SD value of MBSG is the closest to the reference SD (−0.6% difference versus −2.3% of MLP4). Model performance also varies across different water year types. Highest R values of all four models occur in wet years when salinity is generally low. In contrast, R values during dry and critical years (when salinity are normally high on average) are typically the lowest. The smallest and highest RMSD values are observed in below-normal and critical years, respectively, for MBSG, MLP4, and the Hybrid model. For LSTM4, wet years have the smallest RMSD while above-normal years have the highest RMSD. In terms of SD, model performance is generally the worst in critical years, followed by dry years. On average, above-normal years have the most satisfactory SD values.

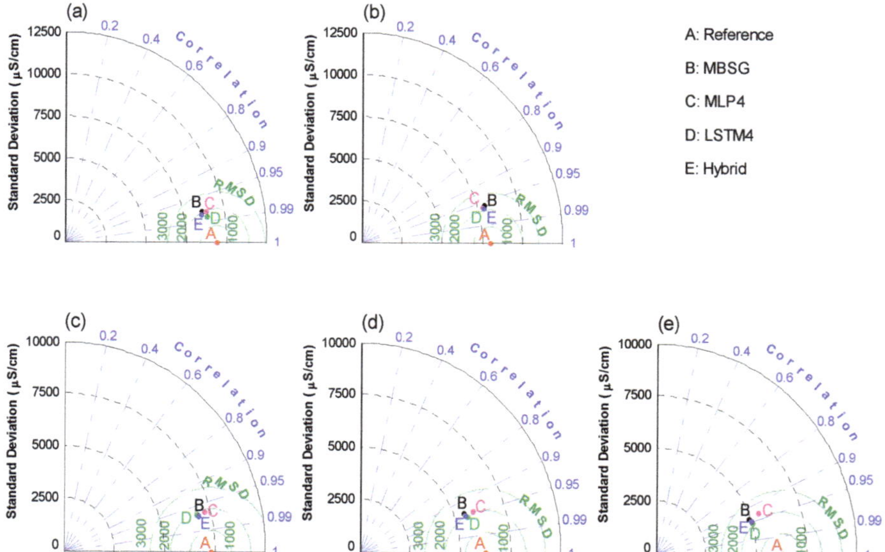

Figure 7. Taylor Diagrams displaying statistics (correlation, standard deviation, and centered root mean square difference) between the reference salinity at Martinez and the corresponding salinity simulations generated via four different models (MLP4, LSTM4, and Hybrid MBSG–CNN) grouped in five water year types including (**a**) wet year, (**b**) above-normal year, (**c**) below-normal year, (**d**) dry year, and (**e**) critical year.

The performance of those four models is also compared to that of the remaining models. Table 4 shows the RMSE of all nine neural network models along with the process-based MBSG model. For MLP models, when only NDO and stage data are considered as network input features (MLP1 and MLP2), the resulting RMSE are much larger than the process-based model across all types of water years. Adding MBSG simulations as an additional input (MLP3) largely improves model performance. Fine-tuning hyper-parameters (MLP4) leads to even smaller RMSE in all types of water years except for below-normal years. For LSTM models, when only NDO and average stage are employed (LSTM1), the resulting RMSE values are generally comparable to that of the MBSG. Adding minimum and maximum stage (LSTM2) yields smaller RMSE in general. Incorporating MBSG simulations as input (LSTM3) leads to consistently smaller RMSE values (versus RMSE values of MBSG, LSTM1, and LSTM2) in all five types of water years. Fine-tuning hyper-parameters (LSTM4) does not necessary lead to further improvement. Similar features are also observed in other two metrics (Tables A3 and A4 in the Appendix A). Looking all models together, LSTM3 has the smallest RMSE in wet years; LSTM4 has the smallest RMSE in dry and critical years, while the hybrid model performs the best during above-normal and below-normal years.

Table 4. Centered root mean square difference between reference and simulated Martinez salinity in different types of water years.

Scenarios	Water Year Type				
	Wet (W)	**Above-Normal (AN)**	**Below-Normal (BN)**	**Dry (D)**	**Critical (C)**
MBSG	2028	2277	1836	2146	2323
MLP1	3261	3524	3648	3704	4464
MLP2	3430	3404	3343	3462	3972
MLP3	2041	2223	1870	2062	2331
MLP4	1986	2204	1872	2040	2313
LSTM1	2055	2478	2313	2368	2392
LSTM2	1939	2280	2182	2219	2411
LSTM3	**1603**	2180	1786	2022	2194
LSTM4	1644	2142	1867	**1947**	**2080**
Hybrid	1918	**2110**	**1767**	2031	2169

Bold numbers represent the best metrics.

Similar to what has been observed in the entire evaluation (Figure 4) period and in three sub-periods representing three different salinity ranges (Figure 6), MLP1 and MLP2 tend to be the outlier models with very different bias and MAE from other models (Figure 8). Their MAE values are markedly larger than that of other models. MLP1 considerably over-estimates the salinity in all types of water years except for the critical years, while MLP2 largely under-estimates Martinez salinity in dry and critical years. For the remaining models, the hybrid model performs the best in above-normal years with the smallest bias and MAE; MLP4 is the best performance model in below-normal years; LSTM3 outperforms other models in dry years. For wet years, two models (MLP4 and LSTM3) have the smallest bias and MAE, respectively. For critical years, LSTM3 and LSTM4 has the smallest bias and MAE, respectively.

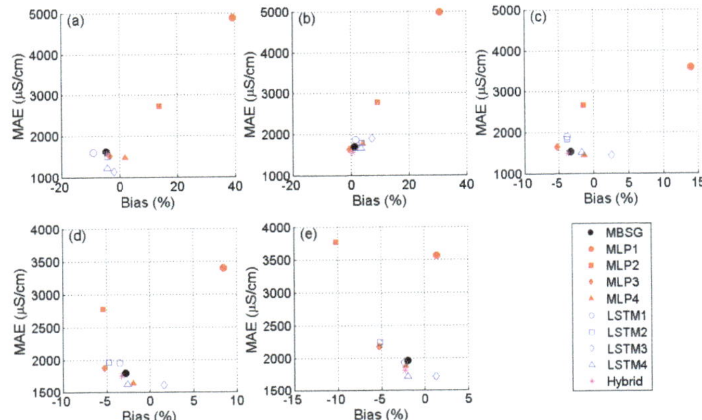

Figure 8. Percent bias (Bias; horizontal axis) and mean absolute error (MAE; vertical axis) between reference and simulated salinity at Martinez via different models grouped in five water year types including (**a**) wet year, (**b**) above-normal year, (**c**) below-normal year, (**d**) dry year, and (**e**) critical year.

Examining five metrics altogether, the neural network models can outperform the process-based MBSG model consistently across all water year types. Table 5 tabulates the improvements calculated as the percent difference between the optimal metrics (of the neural network models with the outlier models MLP1 and MLP2 excluded) and the corresponding MBSG metrics. For R, SD, RMSE, and MAE, the improvements in extreme years (wet, dry, and critical) are more noticeable than the improvements in near-normal (above-normal and below-normal) years. For R and SD (RMSE and MAE), the largest improvements occur in critical (wet) years. The optimal metrics are not associated with a single neural network model. In extreme years, the LSTM models (LSTM3 and LSTM4) tend to be the optimal models; in above-normal years, the hybrid model seems to have the best metrics (except for SD); in below-normal years, the hybrid model and MLP4 perform relatively better in terms of the number of optimal metrics associated with them.

Table 5. Improvements over MBSG metrics in different water year types.

Metrics	Water Year Type				
	W	AN	BN	D	C
R	0.7% (LSTM3/LSTM4)	0.5% (Hybrid)	0.2% (LSTM3/Hybrid)	1.2% (LSTM4)	1.5% (LSTM4)
SD	7.0% (LSTM3)	0.3% (LSTM3)	5.1% (MLP4)	7.0% (MLP4)	8.3% (MLP4)
RMSE	20.9% (LSTM3)	7.3% (Hybrid)	3.8% (Hybrid)	9.3% (LSTM4)	10.5% (LSTM4)
Bias	59.3% (MLP4)	65.3% (Hybrid)	55.1% (MLP4)	45.0% (LSTM3)	33.3% (LSTM3)
MAE	29.1% (LSTM3)	7.7% (Hybrid)	6.1% (MLP4)	10.7% (LSTM3)	12.7% (LSTM4)

4. Discussion

4.1. Data Stationarity and Availability

This study used the first half (water year 1991–2002) of the record period as training/validation period and the second half (water year 2003–2014) as evaluation period. The underlying assumption is that the relationships between salinity and NDO/stage in the first half would hold valid in the second half as well. Put differently, stationarity is assumed in the data employed. To validate this assumption, trend assessment is conducted for mean, maximum, and minimum NDO, Martinez salinity and stage variables on both annual and monthly scales. The widely used non-parametric Mann–Kendall test [70,71] is used in assessing the significance of trend in these variables with a significance level of 0.05. The slope of a significant trend is determined via the Theil–Sen approach [72,73].

Figure 9 depicts the significance level (*p*-value) of the trends in these variables. There is no statistically significant trend in salinity on annual scale or monthly scale. This is also the case for NDO with one exception; mean NDO in February has a significant decreasing trend (-4.4 million m^3/year). Similarly, there is generally no statistically trend in minimum Martinez stage with one exception in March (slope = -0.08 cm/year). For maximum stage, however, significant upward trends are identified in seven out of 12 months. The trend slopes in these months range from 0.06 cm/year (December) to 0.12 cm/year (June). On an annual scale, the trend is also significant with a slope of 0.08 cm/year. Compared to maximum stage, mean stage tends to have significant trends in most months except for January and April. The difference is that, there are downward trends in February and March (both at a rate of -0.04 cm/year). Upward trends in other months with significant trends are generally milder in slope, ranging from 0.03 cm/year to 0.06 cm/year. The trend slope is also milder on annual scale with a value of 0.01 cm/year. This upward trend in mean stage at Martinez is likely linked to the mean sea level rise recorded at the Golden Gate [74].

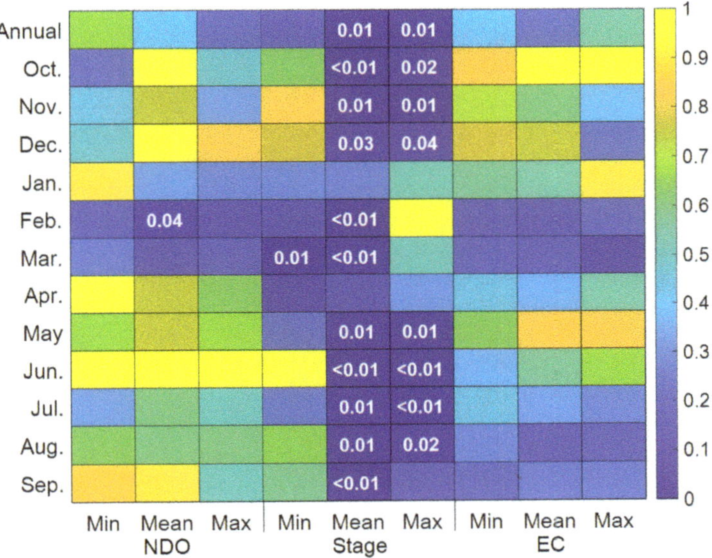

Figure 9. *p*-values of monotonic trends in minimum (Min), mean (Mean), and maximum (Max), net Delta outflow (NDO), Martinez stage (Stage), and salinity (EC) identified via the Mann–Kendall trend assessment on annual scale and monthly scale. A value less than 0.05 indicates that the trend is statistically significant. The *p*-values associated with significant trends are specified.

In this study, Martinez salinity is the predictand while Martinez stage and NDO are predictors. As previously shown in Figure 2, NDO is the primary predictor (R = −0.91) while the mean stage (R = 0.11) is the secondary predictor. Section 3 shows that adding minimum and maximum stage information as additional input features leads to marginal improvement in neural network model performance, suggesting that minimum stage and maximum stage are also minor predictors. Figure 9 illustrates that there are generally no statistically significant trends in the predictand and the primary predictor. Figure 9 also shows that the slopes of significant trend in mean and maximum Martinez stage are mostly mild. In the entire evaluation period (2003–2014), for instance, the overall increase in mean stage amounts to about 1.2 mm (at an annual rate of 0.01 cm/year). This change in the secondary predictor should have minimal impact on the predictability of the predictand.

Nevertheless, sea level rise near Golden Gate is expected to accelerate in pace in the future [75]. Consequently, Martinez stage likely increases at a higher rate. Its influence on Martinez salinity would continue to grow till becoming non-trivial. That poses challenges to process-based models in reliably simulating Martinez salinity, as those models are typically calibrated based on historical conditions which would not be representative for future conditions anymore. Under these circumstances, neural network models have the advantage of learning the trend embedded in the data and applying it into the projections.

It is worth noting that the deep learning methods proposed in the current study use only a subset of the data available in training and predicting. Specifically, the long-term (water year 1991–2014) salinity and stage data available is at 15 min time step. The deep learning methods utilize daily data (aggregated from 15 min data) which only accounts for about 1% of the original salinity and stage data. Nevertheless, the deep learning methods mostly yield superior results when compared to the process-based model. This highlights the robustness of the deep learning methods. This type of method has the potential to be applied to other riverine or estuarine environments where observations are temporally limited, given that the observations are on the key predictors of the predictand.

When observations are spatially limited, model simulations can serve as a viable option in developing and applying machine learning (including deep learning) models [38].

4.2. Estimation of High Range Salinity

This study examines five statistical metrics when evaluating the performance of proposed neural network models against that of the process-based benchmark model. The values of these metrics are solid overall, reflecting satisfactory performance of the proposed deep learning models. However, it is noticeable that the value of one metric is only fair. Specifically, the correlation coefficient between simulated high salinity and the corresponding reference high salinity is remarkably lower than its counterparts for low and medium ranges of salinity (Figure 5c; Table 3). For the benchmark model, the correlation value associated with high range salinity is 0.539, which is much smaller than that of the low range salinity (0.843) and medium range salinity (0.847). The highest correlation of high range salinity simulations is 0.599 (of the hybrid model). LSTM3 (0.598) and LSTM4 (0.589) also yield higher correlation values than the benchmark model. In comparison, the best correlation metrics for low and medium range salinity are 0.883 (LSTM4) and 0.863 (LSTM4), respectively. These observations indicate that: (1) all models are relatively poorer in simulating the variability in high (versus low or medium) range salinity; (2) even though neural network models show improvements over the benchmark model, the corresponding simulations still explain less than 36% (versus 29% of the benchmark model in terms of R^2 from simple linear regressions) of the variability in the reference high range salinity. Additionally, all models (except for the outlier model MLP1) tend to under-estimate high range salinity (Figure 6c). This negative bias is also evident when looking at the exceedance probability curves of the reference and modeled salinity together (Figure A3).

Three additional neural network models are developed to explore the possibility of better modeling the variability and reducing model bias in high range salinity. These models are based on LSTM4 which yields the most favorable R, MAE, RMSD and near-optimal bias and SD during the entire evaluation period (Figures 3 and 4). The first model (LSTM4/D120) differs from LSTM4 in that it uses data from the previous 120 (rather than 60) days to generate next day's salinity. The presumption is that a longer dataset may add new information for the model to learn and predict. The second model (LSTM4/Weight) applies a higher weight to high range salinity in the loss function, while LSTM4 utilizes equal weights to all ranges of flows in the loss function. The expectation is that the model prioritizes high range salinity over low–medium range salinity in its learning and predicting process. The third model (LSTM4/SL) incorporates mean daily sea level observations near Golden Gate as an additional model input. The hypothesis is that as a surrogate of the original salinity source of the Delta, sea level may add information which is not contained in water stage at Martinez. Results of these models are illustrated in Figure 10, along with that of the benchmark model, LSTM4, and the hybrid model.

Looking at the probability exceedance curves of the high range salinity (Figure 10a), the benchmark model, LSTM4, and the hybrid model all under-estimate the reference salinity, so do LSTM4/Weight and LSTM4/SL. This is particularly true at the lower end of the high salinity range (with an exceedance probability over 80%). For LSTM4/D120, however, the exceedance curve becomes remarkably closer to that of the reference salinity. Nevertheless, when looking at the corresponding time series (the insert figure of Figure 10a), a negative bias is still noticeable. The model (i.e., LSTM4/D120) does not capture the variability in reference high range salinity well. These are also reflected in the bias (Figure 10b) and correlation (Figure 10c) plots. The overall bias of LSTM4/D120 in high range salinity is the smallest among all models. Contrariwise, LSTM4/D120 has the largest bias in other salinity ranges, especially for low range salinity (bias = −57.1%). This suggests that LSTM4/D120 reduces bias for high range salinity at the expense of increasing bias for low–medium range salinity. Additionally, LSTM4/D120 has the smallest correlation coefficients across all ranges of salinity (Figure 10c). Except for LSTM/D120, two other proposed models (LSTM4/Weight and LSTM4/SL) have similar biases as LSTM4 (Figure 10a,b) in high range salinity. However, LSTM4/Weight has higher bias in low range salinity while LSTM4/SL has higher bias in medium range salinity compared to LSTM4 (Figure 10b). In terms of correlation,

LSTM4/SL has the most favorable value (R = 0.614) in high range salinity, while LSTM/Weight yields no improvement over LSTM4 nor the hybrid model. For low and medium ranges of salinity, however, LSTM4/SL has smaller correlation coefficients compared to LSTM4 or the hybrid model. In addition, for the entire salinity range, LSTM4 still has the highest correlation value among all models.

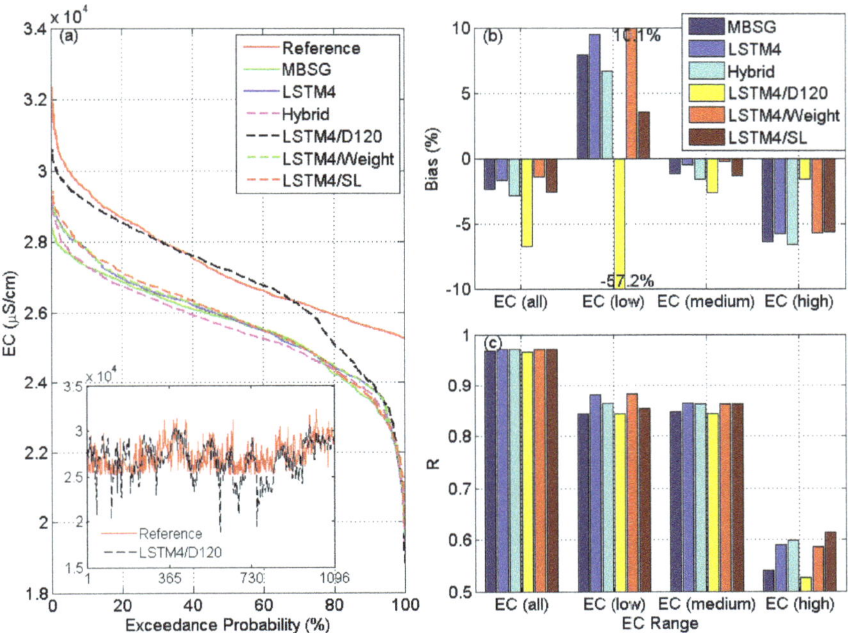

Figure 10. (a) Exceedance probability curves of the reference high range salinity at Martinez and corresponding simulations generated via MBSG, LSTM4, hybrid model, and three additional LSTM models (LSTM/D120, LSTM4/Weight, and LSTM4/SL). The insert figure shows time series plot of the reference high range salinity and the corresponding simulations from LSTM4/D120. (b) Correlation between reference salinity and simulated salinity via different models shown in (a). (c) Percent bias between reference and simulated salinity across different salinity ranges (low, medium, high, and all together).

In brief, out of three additional neural network models proposed, LSTM4/D120 yields less bias and LSTM4/SL slightly improves estimation on the variability of the high range salinity compared to the benchmark MBSG model, LSTM4, and the hybrid model. However, these improvements in high range salinity are at the cost of deteriorated model performance in low and medium ranges of salinity. Further research is warranted to better model high range salinity without compromising on model performance in low–medium range salinity.

4.3. Implications and Future Work

The findings of this study have both scientific and practical implications. From a scientific perspective, the study demonstrates the feasibility of state-of-the-art deep learning techniques in salinity estimation in the Sacramento–San Joaquin Delta (Delta) for the first time. Traditional multilayer perceptron (MLP) networks have been developed and successfully applied in estimating Delta salinity [34,37,39]. This study shows that, when driven by the same NDO and Martinez stage input features, deep learning neural networks (e.g., LSTM1 and LSTM2) distinctly outperform the classic MLP networks (e.g., MLP1 and MLP2) in estimating different ranges of salinity at Martinez and across different water year types. The study further shows that, when trained and validated using only half (versus 85% in those previous MLP studies) of the dataset in the record period, deep learning models (e.g., LSTM3 and LSTM4) can outperform the well-calibrated process-based model (i.e., PEST-calibrated MBSG). These findings lay foundation for developing more sophisticated and carefully designed deep learning architectures to further improve salinity (particularly high range salinity as discussed in Section 4.2) estimation in the Delta. For instance, previous studies have shown that the general ability of artificial neural networks (ANNs) can be improved by combining several ANNs in an ensemble [76–79]. This study indicates that different deep learning neural network models exhibit different strengths in modeling different ranges of salinity across different water year types. Combining the strengths of different models is expected to yield better performance than using individual models alone. This can be achieved by assigning a weight to the output of each model. The weights can be determined from different methods ranging from simple averaging to more complicated Bayesian methods [80]. As another example, the hybrid model examined in this study applies the one-dimensional convolutional neural network (Conv1D) to recognize abstract patterns in dense stage observations which may be ignored by the process-based model (Equation (5)). As indicated previously (Section 2.1; Section 4.1), net Delta outflow (NDO) is the primary predictor for Martinez salinity while stage is the secondary predictor. Presumably applying Conv1D directly to NDO (versus stage) should yield even better performance. Both fronts (i.e., multi-model ensemble and Conv1D configuration for NDO) will be explored in our future work.

From a practical perspective, the PEST-calibrated Martinez Boundary Salinity Generator (MBSG) is mainly applied in generating downstream boundary salinity for the hydrodynamics and water quality model Delta Simulation Model II (DSM2). DSM2 is the operational model used in guiding real-time State Water Project (SWP) and Central Valley Project (CVP) operations and long-term Delta planning studies ranging from climate change, water system operations, to assessment of impacts of potential physical changes in the Delta (dredging, subsidence, island flooding, new water infrastructure, etc.) [81]. The deep learning models (e.g., LSTM3, LSTM4, and the hybrid model) developed in this study have the potential to supplement MBSG for this purpose. In addition, DSM2 currently simulates salinity at 90 locations including those water quality compliance locations in the Delta. The deep learning models developed in this study can be extended to emulate DSM2 in simulating salinity at all these locations. These models, once trained and validated, are expected to run much faster than DSM2. This is very meaningful and particularly appealing to time-sensitive (i.e., real-time) operations in the Delta. These models are also more flexible in terms of requiring less input data. For example, DSM2 needs channel geometry data to accurately simulate flow conditions based on which salinity is derived. The deep learning models, in comparison, does not necessarily need such input and can learn from in-situ flow observations directly. As another example, the current operational version of DSM2 uses Martinez salinity as its downstream salinity boundary as Martinez serves as the physical downstream boundary of the model, while Martinez is not a salinity source and its salinity level is dominated by salty tides from the Pacific Ocean. This study illustrates that the deep learning model can be adapted to directly include sea level as an additional model input (i.e., LSTM4/SL in Section 4.2). The results compare favorably to that of other deep learning models proposed earlier which have been shown to outperform the benchmark MBSG (Figure 10). This flexibility makes deep learning models distinctly attractive to long-term planning studies as sea level rise is projected to be a growing stress to the Delta as well as water operations in the Delta [82,83]. Emulation of DSM2 via deep learning with sea level as an additional input is ongoing and will be reported in our future work.

5. Conclusions

This study aims to explore the potential of deep learning techniques in emulating the process-based Martinez Boundary Salinity Generator (MBSG) in simulating downstream salinity boundary for the Sacramento–San Joaquin Delta of California, United States. The calibrated MBSG is used as the benchmark model. Results indicate that deep learning neural networks are able to provide Martinez salinity simulations that are competitive or superior compared with the benchmark model, particularly when the output of the latter are incorporated as an input to the former. The improvements are generally more noticeable during extreme (wet, dry, and critical) years rather than in near-normal (above-normal and below-normal) years and during low and medium ranges of salinity rather than high range salinity. In a nutshell, this study indicates that deep learning approaches have the potential to supplement the current practices in estimating salinity at Martinez and other locations across the Delta, and thus guide real-time operations and long-term planning activities in the Delta.

Author Contributions: Conceptualization, M.H., L.Z., P.S., and Y.Z.; Data curation, L.Z. and Y.Z.; Funding acquisition, P.S.; Investigation, M.H. and Y.Z.; Methodology, M.H., L.Z., and P.S.; Project administration, P.S.; Software, L.Z.; Validation, M.H.; Visualization, M.H.; Writing—original draft, M.H.; Writing—review and editing, L.Z., P.S., and Y.Z. All authors have read and agreed to the published version of the manuscript.

Funding: This research received no external funding.

Acknowledgments: The authors would like to thank their colleagues Tariq Kadir, Norman Johns, and Eli Ateljevich for reviewing an earlier version of the paper. The authors would also like to thank two anonymous reviewers for providing valuable comments that helped improve the quality of the paper. Furthermore, the authors want to thank Tara Smith for her management support on the study. The views expressed in this paper are those of the authors, and not of the State of California.

Conflicts of Interest: The authors declare no conflict of interest.

Appendix A

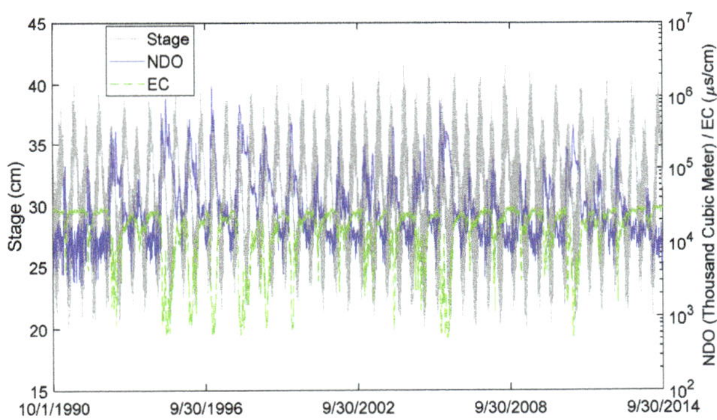

Figure A1. Daily Martinez water stage (Stage), net Delta outflow (NDO) and Martinez salinity represented by electrical conductivity (EC) during the study period (water year 1991–2014).

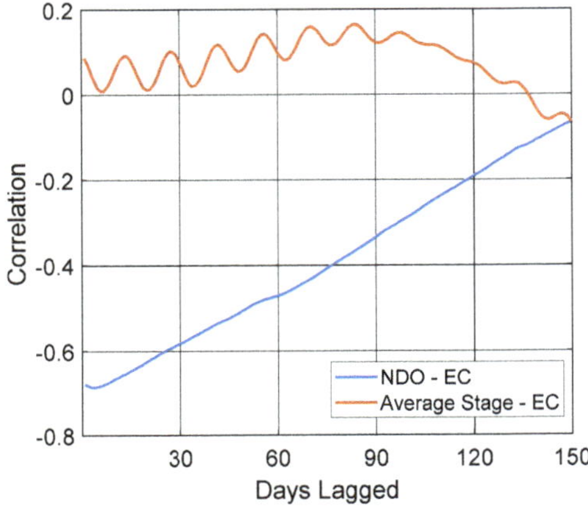

Figure A2. Correlation between daily Martinez salinity and lagged Martinez stage as well as net Delta outflow.

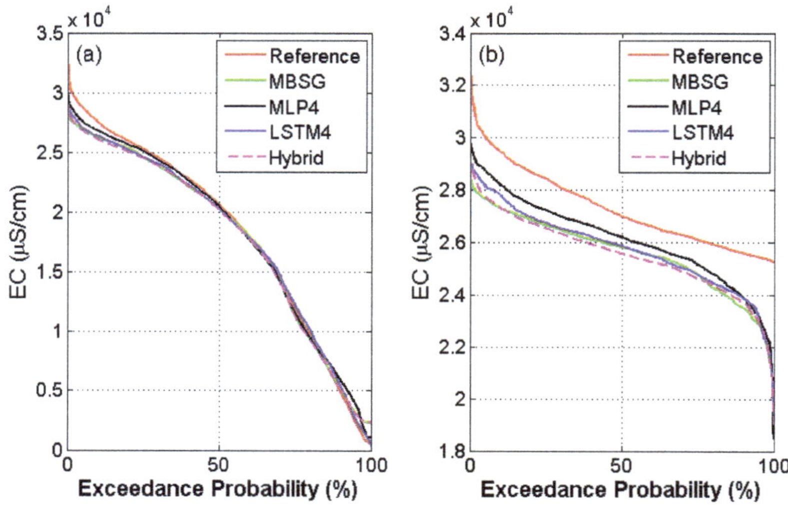

Figure A3. Exceedance probability curves of reference and simulated (via MBSG, MLP4, LSTM4, and the hybrid model) (**a**) salinity and (**b**) high range only salinity during the evaluation period from water year 2003 to 2014.

Table A1. The maximum 14 day running average of mean daily salinity (EC, μs/cm) at Emmaton and Jersey Point in different types of water year [20].

	Emmaton		Jersey Point	
Water Year Type	**450 EC from 1 April to Date Shown**	**EC Value from Date Shown to 15 August**	**450 EC from 1 April to Date Shown**	**EC Value from Date Shown to 15 August**
Wet (W)	15 August	-	15 August	-
Above-Normal (AN)	1 July	630	15 August	-
Below-Normal (BN)	20 June	1140	20 June	740
Dry (D)	15 June	1670	15 June	1350
Critical (C)	-	2780	-	2200

Table A2. Hyper-parameters of the proposed MLP and LSTM models.

Models	Number of Hidden Layers	Number of Neurons (for MLP) or Filters (for LSTM)	Batch Size	Initial Learning Rate	Dropout between Layers	Dropout between Time Steps
MLP1	2	16 and 4	128	1×10^{-3}	-	-
MLP2	2	16 and 4	64	1×10^{-3}	-	-
MLP3	2	32 and 2	128	1×10^{-3}	-	-
MLP4	2	32 and 2	128	1×10^{-6}	-	-
LSTM1	2	64 and 8	256	1×10^{-2}	0	25%
LSTM2	2	8 and 32	64	1×10^{-2}	0	25%
LSTM3	3	32, 32, and 64	128	1×10^{-2}	0	0
LSTM4	3	32, 32, and 64	128	1×10^{-6}	0	0

Table A3. Percent differences (%) between the standard deviation of reference salinity and simulated salinity at Martinez in different types of water years.

Scenarios	Water Year Type				
	W	**AN**	**BN**	**D**	**C**
MBSG	−7.5	−0.6	−6.2	−12.1	−21.5
MLP1	−1.4	−17.5	−1.2	5.6	**−1.0**
MLP2	−5.4	−21.0	−4.9	**1.1**	−2.9
MLP3	−6.2	−2.7	−1.6	−5.3	−13.2
MLP4	−5.0	−2.3	**−1.1**	−5.1	−13.2
LSTM1	−2.3	0.9	−3.6	−7.5	−17.9
LSTM2	−9.8	−8.8	−11.4	−12.9	−23.4
LSTM3	**−0.5**	**0.3**	−4.9	−11.4	−21.3
LSTM4	−5.0	−2.2	−7.0	−10.7	−19.7
Hybrid	−8.6	−1.6	−5.7	−11.7	−20.4

Bold numbers represent the best metrics.

Table A4. Correlation between reference and simulated Martinez salinity in different types of water years.

Scenarios	Water Year Type				
	W	**AN**	**BN**	**D**	**C**
MBSG	0.978	0.959	0.968	0.941	0.941
MLP1	0.938	0.900	0.873	0.834	0.741
MLP2	0.930	0.912	0.890	0.847	0.792
MLP3	0.977	0.960	0.967	0.944	0.930
MLP4	0.977	0.961	0.967	0.945	0.931
LSTM1	0.975	0.952	0.948	0.925	0.930
LSTM2	0.981	0.959	0.957	0.937	0.938
LSTM3	**0.985**	0.963	**0.970**	0.948	0.950
LSTM4	**0.985**	0.963	0.967	**0.952**	**0.955**
Hybrid	0.981	**0.964**	**0.970**	0.948	0.950

Bold numbers represent the best metrics.

References

1. Alber, M. A conceptual model of estuarine inflow policy. *Estuaries* **2002**, *25*, 1246–1261. [CrossRef]
2. Jassby, A.D.; Kimmerer, W.J.; Monismith, S.G.; Armor, C.; Cloern, J.E.; Powell, T.M.; Schubel, J.R.; Vendlinski, T.J. Isohaline position as a habitat indicator for estuarine populations. *Ecol. Appl.* **1995**, *5*, 272–289. [CrossRef]
3. Feyrer, F.; Hobbs, J.; Sommer, T. Salinity inhabited by age-0 splittail (Pogonichthys macrolepidotus) as determined by direct field observation and retrospective analyses with otolith chemistry. *San Franc. Estuary Watershed Sci.* **2010**, *10*. [CrossRef]
4. Moyle, P.B.; Lund, J.R.; Bennett, W.A.; Fleenor, W.E. Habitat variability and complexity in the upper San Francisco Estuary. *San Franc. Estuary Watershed Sci.* **2010**, *24*. [CrossRef]
5. Howard, R.J.; Mendelssohn, I.A. Salinity as a constraint on growth of oligohaline marsh macrophytes. I. Species variation in stress tolerance. *Am. J. Bot.* **1999**, *86*, 785–794. [CrossRef]
6. Weilhoefer, C.L.; Nelson, W.G.; Clinton, P.; Beugli, D.M. Environmental determinants of emergent macrophyte vegetation in Pacific Northwest estuarine tidal wetlands. *Estuaries Coasts* **2013**, *36*, 377–389. [CrossRef]
7. Borgnis, E.; Boyer, K.E. Salinity tolerance and competition drive distributions of native and invasive submerged aquatic vegetation in the Upper San Francisco Estuary. *Estuaries Coasts* **2016**, *39*, 707–717. [CrossRef]

8. Fernández-Delgado, C.; Baldó, F.; Vilas, C.; García-González, D.; Cuesta, J.; González-Ortegón, E.; Drake, P. Effects of the river discharge management on the nursery function of the Guadalquivir river estuary (SW Spain). *Hydrobiologia* **2007**, *587*, 125–136. [CrossRef]

9. Wang, Y.; Jiao, J.J. Origin of groundwater salinity and hydrogeochemical processes in the confined Quaternary aquifer of the Pearl River Delta, China. *J. Hydrol.* **2012**, *438*, 112–124. [CrossRef]

10. Smajgl, A.; Toan, T.Q.; Nhan, D.K.; Ward, J.; Trung, N.H.; Tri, L.; Tri, V.; Vu, P. Responding to rising sea levels in the Mekong Delta. *Nat. Clim. Chang.* **2015**, *5*, 167–174. [CrossRef]

11. Hutton, P.H.; Rath, J.S.; Chen, L.; Ungs, M.J.; Roy, S.B. Nine decades of salinity observations in the San Francisco Bay and Delta: Modeling and trend evaluations. *J. Water Resour. Plan. Manag.* **2016**, *142*, 04015069. [CrossRef]

12. Johns, N.D.; Heger, N.A. Characterizing estuarine salinity patterns with event duration and frequency of reoccurrence approaches. *Limnol. Oceanogr.-Meth.* **2018**, *16*, 180–198. [CrossRef]

13. United States Census Bureau. United States Census Bureau QuickFacts: United States. 2019. Available online: https://www.census.gov/quickfacts/fact/table/US/PST045219 (accessed on 1 May 2020).

14. Sabet, M.H.; Coe, J.Q. Models for water and power scheduling for the California State Water Project. *J. Am. Water Resours.* **1986**, *22*, 587–596. [CrossRef]

15. Becker, L.; Yeh, W.; Fults, D.; Sparks, D. Operations models for central valley project. *J. Water Resour. Plann. Manag. Div. Am. Soc. Civ. Eng.* **1976**, *102*, 101–115.

16. Myers, N.; Mittermeier, R.A.; Mittermeier, C.G.; Da Fonseca, G.A.; Kent, J. Biodiversity hotspots for conservation priorities. *Nature* **2000**, *403*, 853. [CrossRef]

17. Moyle, P.B.; Brown, L.R.; Durand, J.R.; Hobbs, J.A. Delta smelt: Life history and decline of a once-abundant species in the San Francisco Estuary. *San Franc. Estuary Watershed Sci.* **2016**, *30*. [CrossRef]

18. Healey, M.; Dettinger, M.; Norgaard, R. Perspectives on Bay–Delta Science and Policy. *San Franc. Estuary Watershed Sci.* **2016**, *25*.

19. DSC. *The Delta Plan*; Delta Stewardship Counci: Sacramento, CA, USA, 2013.

20. CSWRCB. *Water right Decision 1641*; CSWRCB: Sacramento, CA, USA, 1999; p. 225.

21. USFWS. *Formal Endangered Species Act Consultation on the Proposed Coordinated Operations of the Central Valley Project (CVP) and State Water Project (SWP)*; USFWS: Sacramento, CA, USA, 2008; p. 410.

22. Denton, R.A. Accounting for antecedent conditions in seawater intrusion modeling—Applications for the San Francisco Bay-Delta. In *Hydraulic Engineering*; CSWRCB: Sacramento, CA, USA, 1993; pp. 448–453.

23. CDWR. Calibration and verification of DWRDSM. In *Methodology for Flow and Salinity Estimates in the Sacramento-San Joaquin Delta and Suisun Marsh: 12th Annual Progress Report*; CDWR: Sacramento, CA, USA, 1991.

24. DeGeorge, J.F. *A Multi-Dimensional Finite Element Transport Model Utilizing a Characteristic-Galerkin Algorithm*; University of California: Davis, CA, USA, 1996.

25. Cheng, R.T.; Casulli, V.; Gartner, J.W. Tidal, residual, intertidal mudflat (TRIM) model and its applications to San Francisco Bay, California. *Estuar. Coast. Shelf Sci.* **1993**, *36*, 235–280. [CrossRef]

26. CDWR. Bay-Delta SCHISM Model Developments and Applications. In *Methodology for Flow and Salinity Estimates in the Sacramento-San Joaquin Delta and Suisun Marsh: 36th Annual Progress Report*; CDWR: Sacramento, CA, USA, 2015.

27. Chao, Y.; Farrara, J.D.; Zhang, H.; Zhang, Y.J.; Ateljevich, E.; Chai, F.; Davis, C.O.; Dugdale, R.; Wilkerson, F. Development, implementation, and validation of a modeling system for the San Francisco Bay and Estuary. *Estuar. Coast. Shelf Sci.* **2017**, *194*, 40–56. [CrossRef]

28. Casulli, V.; Zanolli, P. Semi-implicit numerical modeling of nonhydrostatic free-surface flows for environmental problems. *Math. Comput. Model* **2002**, *36*, 1131–1149. [CrossRef]

29. MacWilliams, M.; Bever, A.J.; Foresman, E. 3-D simulations of the San Francisco Estuary with subgrid bathymetry to explore long-term trends in salinity distribution and fish abundance. *San Franc. Estuary Watershed Sci.* **2016**, *24*. [CrossRef]

30. Fringer, O.; Gerritsen, M.; Street, R. An unstructured-grid, finite-volume, nonhydrostatic, parallel coastal ocean simulator. *Ocean Model.* **2006**, *14*, 139–173. [CrossRef]

31. Hsu, K.; Stacey, M.T.; Holleman, R.C. Exchange between an estuary and an intertidal marsh and slough. *Estuaries Coasts* **2013**, *36*, 1137–1149. [CrossRef]

32. MacWilliams, M.L.; Ateljevich, E.S.; Monismith, S.G.; Enright, C. An Overview of Multi-Dimensional Models of the Sacramento–San Joaquin Delta. *San Franc. Estuary Watershed Sci.* **2016**, *35*.

33. Ateljevich, E. Improving Estimates of Salinity at the Martinez Boundary. In *Methodology for Flow and Salinity Estimates in the Sacramento-San Joaquin Delta and Suisun Marsh: 22nd Annual Progress Report*; CDWR: Sacramento, CA, USA, 2001.

34. Finch, R.; Sandhu, N. *Artificial Neural Networks with Application to the Sacramento–San Joaquin Delta*; California Department of Water Resources Delta Modeling Section, Division of Planning: Sacramento, CA, USA, 1995.

35. CDWR. Integration of CALSIM and Artificial Neural Networks Models for Sacramento-San Joaquin Delta Flow-Salinity Relationships. In *Methodology for Flow and Salinity Estimates in the Sacramento-San Joaquin Delta and Suisun Marsh: 22nd Annual Progress Report*; CDWR: Sacramento, CA, USA, 2001.

36. Chung, F.I.; Seneviratne, S.A. Developing artificial neural networks to represent salinity intrusions in the Delta. In Proceedings of the World Environmental and Water Resources Congress 2009: Great Rivers, Kansas City, MO, USA, 17–21 May 2009; pp. 1–10.

37. Rath, J.S.; Hutton, P.H.; Chen, L.; Roy, S.B. A hybrid empirical-Bayesian artificial neural network model of salinity in the San Francisco Bay-Delta estuary. *Environ. Model. Softw.* **2017**, *93*, 193–208. [CrossRef]

38. Chen, L.; Roy, S.B.; Hutton, P.H. Emulation of a process-based estuarine hydrodynamic model. *Hydrol. Sci. J.* **2018**, *63*, 783–802. [CrossRef]

39. Jayasundara, N.C.; Seneviratne, S.A.; Reyes, E.; Chung, F.I. Artificial Neural Network for Sacramento–San Joaquin Delta Flow–Salinity Relationship for CalSim 3.0. *J. Water Resour. Plan. Manag.* **2020**, *146*, 04020015. [CrossRef]

40. Maier, H.R.; Jain, A.; Dandy, G.C.; Sudheer, K.P. Methods used for the development of neural networks for the prediction of water resource variables in river systems: Current status and future directions. *Environ. Model. Softw.* **2010**, *25*, 891–909. [CrossRef]

41. Rumelhart, D.E.; Hinton, G.E.; Williams, R.J. Learning representations by back-propagating errors. *Nature* **1986**, *323*, 533–536. [CrossRef]

42. Shen, C. A transdisciplinary review of deep learning research and its relevance for water resources scientists. *Water Resour. Res.* **2018**, *54*, 8558–8593. [CrossRef]

43. Elman, J.L. Finding structure in time. *Cogn. Sci.* **1990**, *14*, 179–211. [CrossRef]

44. Bengio, Y.; Simard, P.; Frasconi, P. Learning long-term dependencies with gradient descent is difficult. *IEEE Trans. Neural Netw.* **1994**, *5*, 157–166. [CrossRef]

45. Hochreiter, S.; Schmidhuber, J. Long short-term memory. *Neural Comput.* **1997**, *9*, 1735–1780. [CrossRef]

46. Kratzert, F.; Klotz, D.; Brenner, C.; Schulz, K.; Herrnegger, M. Rainfall–runoff modelling using long short-term memory (LSTM) networks. *Hydrol. Earth Syst. Sci.* **2018**, *22*, 6005–6022. [CrossRef]

47. Gu, H.; Xu, Y.-P.; Ma, D.; Xie, J.; Liu, L.; Bai, Z. A surrogate model for the Variable Infiltration Capacity model using deep learning artificial neural network. *J. Hydrol.* **2020**, 125019. [CrossRef]

48. Tian, Y.; Xu, Y.-P.; Yang, Z.; Wang, G.; Zhu, Q. Integration of a parsimonious hydrological model with recurrent neural networks for improved streamflow forecasting. *Water* **2018**, *10*, 1655. [CrossRef]

49. Fan, H.; Jiang, M.; Xu, L.; Zhu, H.; Cheng, J.; Jiang, J. Comparison of Long Short Term Memory Networks and the Hydrological Model in Runoff Simulation. *Water* **2020**, *12*, 175. [CrossRef]

50. Zhang, J.; Zhu, Y.; Zhang, X.; Ye, M.; Yang, J. Developing a Long Short-Term Memory (LSTM) based model for predicting water table depth in agricultural areas. *J. Hydrol.* **2018**, *561*, 918–929. [CrossRef]

51. Bowes, B.D.; Sadler, J.M.; Morsy, M.M.; Behl, M.; Goodall, J.L. Forecasting Groundwater Table in a Flood Prone Coastal City with Long Short-term Memory and Recurrent Neural Networks. *Water* **2019**, *11*, 1098. [CrossRef]

52. Liang, C.; Li, H.; Lei, M.; Du, Q. Dongting lake water level forecast and its relationship with the three gorges dam based on a long short-term memory network. *Water* **2018**, *10*, 1389. [CrossRef]

53. Yan, J.; Chen, X.; Yu, Y.; Zhang, X. Application of a Parallel Particle Swarm Optimization-Long Short Term Memory Model to Improve Water Quality Data. *Water* **2019**, *11*, 1317. [CrossRef]

54. Zhou, J.; Wang, Y.; Xiao, F.; Wang, Y.; Sun, L. Water Quality Prediction Method Based on IGRA and LSTM. *Water* **2018**, *10*, 1148. [CrossRef]

55. Zhang, D.; Peng, Q.; Lin, J.; Wang, D.; Liu, X.; Zhuang, J. Simulating Reservoir Operation Using a Recurrent Neural Network Algorithm. *Water* **2019**, *11*, 865. [CrossRef]

56. Krizhevsky, A.; Sutskever, I.; Hinton, G.E. Imagenet classification with deep convolutional neural networks. In *Advances in Neural Information Processing Systems*; NIPS: Lake Tahoe, CA, USA, 2012; pp. 1097–1105.

57. Chen, Y.; Li, W.; Sakaridis, C.; Dai, D.; Van Gool, L. Domain adaptive faster r-cnn for object detection in the wild. In Proceedings of the IEEE Conference on Computer Vision and Pattern Recognition, Salt Lake City, UT, USA, 18–23 June 2018; pp. 3339–3348.

58. Wang, Y.; Sun, Y.; Liu, Z.; Sarma, S.E.; Bronstein, M.M.; Solomon, J.M. Dynamic graph cnn for learning on point clouds. *ACM Trans. Graph.* **2019**, *38*, 1–12. [CrossRef]

59. Zhong, L.; Hu, L.; Zhou, H. Deep learning based multi-temporal crop classification. *Remote Sens. Environ.* **2019**, *221*, 430–443. [CrossRef]

60. Afzaal, H.; Farooque, A.A.; Abbas, F.; Acharya, B.; Esau, T. Groundwater estimation from major physical hydrology components using artificial neural networks and deep learning. *Water* **2020**, *12*, 5. [CrossRef]

61. Pan, B.; Hsu, K.; AghaKouchak, A.; Sorooshian, S. Improving precipitation estimation using convolutional neural network. *Water Resour. Res.* **2019**, *55*, 2301–2321. [CrossRef]

62. Miao, Q.; Pan, B.; Wang, H.; Hsu, K.; Sorooshian, S. Improving monsoon precipitation prediction using combined convolutional and long short term memory neural network. *Water* **2019**, *11*, 977. [CrossRef]

63. Kimura, N.; Yoshinaga, I.; Sekijima, K.; Azechi, I.; Baba, D. Convolutional Neural Network Coupled with a Transfer-Learning Approach for Time-Series Flood Predictions. *Water* **2020**, *12*, 96. [CrossRef]

64. Wang, Y.; Fang, Z.; Hong, H.; Peng, L. Flood susceptibility mapping using convolutional neural network frameworks. *J. Hydrol.* **2020**, *582*, 124482. [CrossRef]

65. Sandhu, P.; Zhou, Y. Calibrating the Martinez Boundary Salinity Generator Using PEST. In *Methodology for Flow and Salinity Estimates in the Sacramento-San Joaquin Delta and Suisun Marsh: 36th Annual Progress Report*; CDWR: Sacramento, CA, USA, 2015.

66. CDWR. *On Estimating Net Delta Outflow (NDO): Approaches to Estimating NDO in the Sacramento-San Joaquin Delta*; CDWR: Sacramento, CA, USA, 2016.

67. Doherty, J.; Johnston, J.M. Methodologies for calibration and predictive analysis of a watershed model. *J. Am. Water Resour. Assoc.* **2003**, *39*, 251. [CrossRef]

68. Kingma, D.P.; Ba, J. Adam: A method for stochastic optimization. *arXiv* **2014**, arXiv:1412.6980.

69. Taylor, K.E. Summarizing multiple aspects of model performance in a single diagram. *J. Geophys. Res. Atmos.* **2001**, *106*, 7183–7192. [CrossRef]

70. Mann, H.B. Nonparametric tests against trend. *Econometrica* **1945**, *13*, 245–259. [CrossRef]

71. Kendall, M. *Rank Correlation Methods*; Charles Griffin: London, UK, 1975.

72. Thiel, H. A rank-invariant method of linear and polynomial regression analysis, Part 3. In Proceedings of the Koninalijke Nederlandse Akademie van Weinenschatpen A, Amsterdam, The Netherlands, 25 February 1950; volume 53, pp. 1397–1412.

73. Sen, P.K. Estimates of the regression coefficient based on Kendall's tau. *J. Am. Stat. Assoc.* **1968**, *63*, 1379–1389. [CrossRef]

74. Cloern, J.E.; Knowles, N.; Brown, L.R.; Cayan, D.; Dettinger, M.D.; Morgan, T.L.; Schoellhamer, D.H.; Stacey, M.T.; Van der Wegen, M.; Wagner, R.W. Projected evolution of California's San Francisco Bay-Delta-River system in a century of climate change. *PLoS ONE* **2011**, *6*, e24465. [CrossRef]

75. Pierce, D.W.; Kalansky, J.F.; Cayan, D.R. *Climate, Drought, and Sea Level Rise Scenarios for California's Fourth Climate Change Assessment*; Technical Report CCCA4-CEC-2018-006; California Energy Commission: Sacramento, CA, USA, 2018.

76. Shu, C.; Burn, D.H. Artificial neural network ensembles and their application in pooled flood frequency analysis. *Water Resour. Res.* **2004**, *40*. [CrossRef]

77. Tiwari, M.K.; Adamowski, J. Urban water demand forecasting and uncertainty assessment using ensemble wavelet-bootstrap-neural network models. *Water Resour. Res.* **2013**, *49*, 6486–6507. [CrossRef]

78. Berkhahn, S.; Fuchs, L.; Neuweiler, I. An ensemble neural network model for real-time prediction of urban floods. *J. Hydrol.* **2019**, *575*, 743–754. [CrossRef]

79. Elkiran, G.; Nourani, V.; Abba, S. Multi-step ahead modelling of river water quality parameters using ensemble artificial intelligence-based approach. *J. Hydrol.* **2019**, *577*, 123962. [CrossRef]

80. Hoeting, J.A.; Madigan, D.; Raftery, A.E.; Volinsky, C.T. Bayesian model averaging: A tutorial. *Stat. Sci.* **1999**, 382–401.

81. Anderson, J.; Chung, F.; Anderson, M.; Brekke, L.; Easton, D.; Ejeta, M.; Peterson, R.; Snyder, R. Progress on incorporating climate change into management of California's water resources. *Clim. Chang.* **2008**, *87*, 91–108. [CrossRef]

82. Dettinger, M.; Anderson, J.; Anderson, M.; Brown, L.R.; Cayan, D.; Maurer, E. Climate change and the Delta. *San Franc. Estuary Watershed Sci.* **2016**, 26. [CrossRef]

83. Ruckert, K.L.; Oddo, P.C.; Keller, K. Impacts of representing sea-level rise uncertainty on future flood risks: An example from San Francisco Bay. *PLoS ONE* **2017**, *12*, e0174666. [CrossRef] [PubMed]

Article

Modelling Study of Transport Time Scales for a Hyper-Tidal Estuary

Guanghai Gao [1,2,*], **Junqiang Xia** [1], **Roger A. Falconer** [3] **and Yingying Wang** [2]

[1]   State Key Laboratory of Water Resources and Hydropower Engineering Science, Wuhan University, Wuhan 430072, China; xiajq@whu.edu.cn

[2]   Tianjin Key Laboratory of Environmental Technology for Complex Trans-Media Pollution, College of Environmental Science and Engineering, Nankai University, Tianjin 300350, China; wangyy@nankai.edu.cn

[3]   Cardiff School of Engineering, Cardiff University, The Parade, Cardiff CF24 3AA, UK; FalconerRA@cardiff.ac.uk

*   Correspondence: gaogh@nankai.edu.cn

Received: 5 July 2020; Accepted: 26 August 2020; Published: 30 August 2020

Abstract: This paper presents a study of two transport timescales (TTS), i.e., the residence time and exposure time, of a hyper-tidal estuary using a widely used numerical model. The numerical model was calibrated against field measured data for various tidal conditions. The model simulated current speeds and directions generally agreed well with the field data. The model was then further developed and applied to study the two transport timescales, namely the exposure time and residence time for the hyper-tidal Severn Estuary. The numerical model predictions showed that the inflow from the River Severn under high flow conditions reduced the residence and exposure times by 1.5 to 3.5% for different tidal ranges and tracer release times. For spring tide conditions, releasing a tracer at high water reduced the residence time and exposure time by 49.0% and 11.9%, respectively, compared to releasing the tracer at low water. For neap tide conditions, releasing at high water reduced the residence time and exposure time by 31.6% and 8.0%, respectively, compared to releasing the tracer at low water level. The return coefficient was found to be vary between 0.75 and 0.88 for the different tidal conditions, which indicates that the returning water effects for different tidal ranges and release times are all relatively high. For all flow and tide conditions, the exposure times were significantly greater than the residence times, which demonstrated that there was a high possibility for water and/or pollutants to re-enter the Severn Estuary after leaving it on an ebb tide. The fractions of water and/or pollutants re-entering the estuary for spring and neap tide conditions were found to be very high, giving 0.75–0.81 for neap tides, and 0.79–0.88 for spring tides. For both the spring and neap tides, the residence and exposure times were lower for high water level release. Spring tide conditions gave significantly lower residence and exposure times. The spatial distribution of exposure and residence times showed that the flow from the River Severn only had a local effect on the upstream part of the estuary, for both the residence and exposure time.

Keywords: residence time; exposure time; transport time scale; hyper-tidal estuary

---

1. Introduction

Coastal waters, such as estuaries, bays etc., play an important role in terms of the transport of receiving wastewater from both anthropogenic and natural sources. These transport processes are affected by various hydrodynamic and environmental factors, leading to complex and dynamic advection and mixing processes in coastal and estuarine water zones. Therefore, a better understanding of the behaviour of the water exchange processes in these water bodies is critical to decision making that underpins our better management of the changing pressures in such hydro-environmental systems. Water exchange processes are the fundamental driving factors governing the transport and fate of various physical, chemical and biological water quality indicators. Transport time scales (TTSs) are the main indexes adopted by water managers and engineers for interpreting the flow in such basins and for describing the effects of advective and dispersive processes on the transport of pollutants in these basins [1]. Various TTSs are reported in the scientific literature to evaluate distinctive aspects of the water exchange processes, such as residence time [2], exposure time [3,4], flushing time [2], e-folding flushing time [2,5], turn over time [5,6], influence time [7] and water age [8]. Recent studies [3,4] have also shown that studying the residence time and exposure time in parallel has the advantage of separating and quantifying the returning water effects on the TTSs, for a controlled domain as a whole and its spatial distribution, while the other TTSs do not have this advantage. The residence time is the time taken for a water parcel, including solutes or particulate matter, to leave a controlled domain for the first time. However, the exposure time is the total time spent by a water parcel and any constituents, in the controlled domain, which includes the time intervals for subsequent re-entries [3,4]. The residence time excludes the time spent by water parcels, including constituents, in the domain following its initial exit from within the domain [4]. This can result in substantial differences between the residence and exposure times, particularly in water bodies where the re-entering of a water parcel has a significant impact on the tidal basin, such as where the water parcel exits from the domain on the ebb tide and then re-enters to the basin on subsequent flood tide. A number of studies have been undertaken to investigate the residence and exposure times together, to acquire a better understanding of a converging shape estuary [3], a reconstructed wetland [9], the micro-tidal Pearl River Estuary [4] and the shallow Dublin Bay with a macro-tidal range. However, there is currently a lack of knowledge to establish the impact of the residence and exposure times in an estuary for hyper-tidal estuaries, where the tidal range is greater than 6 m [10]. Further studies are therefore needed for hyper-tidal estuaries, for both scientific advancement and water quality managerial improvement. The Severn Estuary forms such a hyper-tidal estuary, which has been frequently considered for extracting tidal range power from the basin to supply considerable quantities of renewable energy.

The Severn Estuary is one of the largest estuarine basins in the UK, and is situated in the south west region of the UK, between South East Wales and South West England, as shown in Figure 1. The estuary has one of the highest tidal ranges in the world [11], generating large tidal currents and very high suspended sediment concentrations in excess of 1000 mg/l [12]. The Severn Estuary is a hyper-tidal estuary with typical spring and neap tidal ranges of 13.5 m and 6.5 m respectively [13]. The water exchange processes and transport time scales are important factors in governing sediment transport [14], water quality and the ecosystem of the basin. The combination of high tidal ranges, the funnel shape of the basin and the relatively steep slopes make the findings and conclusions from other estuaries studied uncertain in their applicability to the Severn Estuary. Therefore, this study focused on investigating the residence and exposure times of the Severn Estuary in parallel, to characterise the transport time scales for a hyper-tidal estuary. The effects of freshwater discharges, tidal ranges and the release time of a tracer were considered in computing and predicting the corresponding exposure and residence times for the basin. The spatial distribution of the transport time scales were predicted, in order to identify the effects of the river flows, tide ranges and tracer release times on the TTSs in different regions along the Severn Estuary. The overall return coefficients were also evaluated for various tidal conditions to quantify and evaluate the effects of the returning water parcel on the exchange processes and the transport time scales. The current focus of the transport time scale studies for coastal water bodies has shifted from the global and bulk timescale (i.e., flushing time, turn over time etc.) to the transport time scales, which are more informative and suitable for understanding their spatial distribution [15], such as residence and exposure times. However, there is currently a lack of knowledge on the water exchange processes and TTSs for estuaries such as the Severn Estuary, which forms the focus of this study.

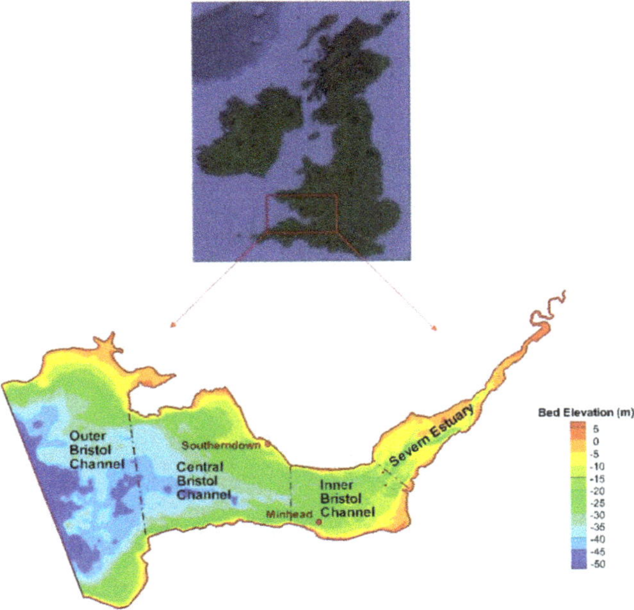

Figure 1. Study area and site location.

2. Materials and Methods

2.1. Transport Time Scale Modelling

The time taken by a water parcel, including constituents, to reach the outlet [16], which means the time taken for a water parcel to leave the control domain for the first time, is defined as the residence time. In this study, the remnant function was adopted, as suggested in various studies [4], to quantify the transport time scales, i.e., residence and exposure time. The remnant function represents the fraction of the initial mass of the tracer remaining in the domain at time t, and is defined as follows:

$$r(t) = \frac{M(t)}{M(t_0)} = \frac{\int_\Omega H(x,y,t) \bullet C(x,y,t)d\Omega}{\int_\Omega H(x,y,t_0) \bullet C(x,y,t_0)d\Omega} \tag{1}$$

where $M(t_0)$ is the total amount of tracer at the initial time t_0, and $M(t)$ is the amount of tracer remaining in the domain after time t; $H(x, y, t)$ = the water depth at location (x, y), time t; $C(x, y, t)$ = the tracer concentration at location (x, y) and time t. The residence time, or exposure time, can then be defined as:

$$T = \int_{t_0}^{+\infty} r(t)dt \tag{2}$$

where T is the residence time or exposure time, and $r(t)$ is the remnant function. The residence time characterises the transport time scale of the estuarine basin, where the water parcel does not return to the basin after reaching the outlet, such as what happens most of the time in rivers and lakes etc. However, in estuarine and coastal zones, where the tide plays a significant role in governing water exchange processes, some of the water parcel returns into the domain after leaving. Hence, the exposure time has been defined to include the returning effects on the transport time scales [3,4,7,17–19]. This approach was therefore adopted in this study.

Both the exposure and residence times in the Severn Estuary were evaluated using a numerical model in this study, where a passive conservative tracer was used as marker to calculate the transport processes in governing domain. The conservative tracer concentration in the interested region, the Severn Estuary (Figure 1), was initially set to 1.0 and 0 elsewhere, as shown in Equation (3). The residence time was determined by counting the time it took to reach the mouth of the estuary for the first time. Therefore, in calculating the residence time, the concentrations were set to zero once the water parcel had reached the mouth of the estuary, as suggested in [4], while the exposure time was calculated, based on including the tracer transported back into the estuary, as summarised in the equations below:

$$C(x, y, t = 0) = \begin{cases} 1 & (x, y) \in \Omega \\ 0 & (x, y) \notin \Omega \end{cases} \tag{3}$$

where Ω = the domain of interest, the Severn Estuary in this study, as shown in Figure 1.

For the investigation of the spatial distribution of the transport time scales, the transport time scales at location (x, y) were calculated as follows:

$$T(x, y) = \int_{t_0}^{+\infty} \frac{H(x, y, t) \bullet C(x, y, t)}{H(x, y, t_0) \bullet C(x, y, t_0)} dt \tag{4}$$

The water exchange processes in the Severn Estuary are highly dynamic and complex, so the residence and the exposure times would be driven and affected by various factors, including the initial release time of the tracer, the tidal range and river flow inputs etc. Therefore, various numerical modelling scenarios and numerical experiments were carried out, to include high and low tidal levels, spring and neap tides and various river flow conditions, to evaluate the effects of these factors on the residence and exposure times.

In a theoretical analysis, the residence and exposure times can be studied by using the method based on integrating the remnant function (Equation (2)) from the initial time (t_0) to infinity ($t_{0+\infty}$), which is impractical for real estuaries. In practice, various studies [3,4] have suggested integrating the remnant function over a finite time period, with this time being sufficiently long enough for most of the tracer to have left the domain of interest.

The return coefficient was suggested [3,4] to quantify the amount of water re-entering the estuary on the transport time scales. This approach was adopted in this study to represent the amount of the water parcel and tracer to the Severn Estuary after leaving the estuary mouth for the first time:

$$C_r = \frac{T_e - T_r}{T_e} \tag{5}$$

C_r is the return coefficient quantifying the contribution of returning water to the exposure time.

2.2. Hydrodynamic and Dispersion Model

The widely used hydro-environmental model Environmental Fluid Dynamics Code (EFDC) [20], was refined and used in this study to simulate the hydrodynamics, and evaluate the residence time, exposure time and return coefficients. The EFDC model uses a boundary-fitted curvilinear grid in the horizontal domain and sigma grids in the vertical direction respectively and has been used in many studies [1,9,20–22]. The governing equations and numerical method used to solve the modelling equations using EFDC have been detailed in various previous publications [1,20,21], and the momentum and continuity equations and transport equations for a conservative tracer are summarised in the following form:

$$\begin{aligned}
\partial_t(mHu) + & \ \partial_x\big(m_yHuu\big) + \partial_y(m_xHvu) + \partial_z(mwu) \\
& -\big(mf + v\partial_xm_y - u\partial_ym_x\big)Hv \\
& = -m_yH\partial_x(g\xi + p) - m_y(\partial_xh - z\partial_xH)\partial_zp \\
& +\partial_z\big(mH^{-1}A_v\partial_zu\big) + Q_u
\end{aligned} \tag{6}$$

$$\begin{aligned}
\partial_t(mHv) + & \ \partial_x\big(m_yHuv\big) + \partial_y(m_xHvv) + \partial_z(mwv) \\
& +\big(mf + v\partial_xm_y - u\partial_ym_x\big)Hu \\
& = -m_xH\partial_y(g\xi + p) - m_x\big(\partial_yh - z\partial_yH\big)\partial_zp \\
& +\partial_z\big(mH^{-1}A_v\partial_zv\big) + Q_v
\end{aligned} \tag{7}$$

$$\partial_zp = -gH(\rho - \rho_0)\rho_0^{-1} = -gHb \tag{8}$$

$$\partial_zp = -gH(\rho - \rho_0) = -gHb$$
$$\partial_t(m\xi) + \partial_x\big(m_yHu\big) + \partial_y(m_xHv) + \partial_z(mw) = 0 \tag{9}$$

$$\partial_t(mHS) + \partial_x\big(m_yHuS\big) + \partial_y(m_xHvS) + \partial_z(mwS)$$
$$= \partial_z\big(mH^{-1}A_b\partial_zS\big) + Q_s \tag{10}$$

where u and v are the horizontal velocity components in the curvilinear plane, x and y are orthogonal coordinates, m_x and m_y are the square roots of the diagonal components of the metric tensor, and $m = m_xm_y$ is the Jacobian or square root of the metric tensor determinant. The total depth $H = h + \xi$, consists of the depth below and the free surface displacement above the undisturbed physical vertical coordinate origin, i.e., $z^* = 0$. In the momentum Equations (6) and (7), f is the Coriolis parameter, A_v is the vertical turbulent or eddy viscosity and Q_u and Q_v are the momentum source-sink terms, which were used to model the subgrid scale horizontal diffusion. The pressure p is the relative hydrostatic pressure in the water column, where ρ and ρ_0 are the actual and reference water densities. $S =$ conservative tracer concentration in the transport equation. The numerical scheme adopted in the EFDC model is based on a combination of the finite volume and finite difference spatial discretisation methods on a C grid staggering of the discrete variables. Full details of the EFDC model are given in the corresponding documents [20].

The bathymetry of the computational domain is shown in Figure 1, where the total model area was approximately 5700 km^2, which covered the whole of the Bristol Channel and the Severn Estuary. The bathymetry used in this model was obtained by interpolation using a digital bathymetric chart of the area downstream of the second Severn Bridge and observed cross-sectional profiles upstream of the bridge and up through the River Severn [12,23]. The model extended a distance of 180 km in the east–west direction and 72 km in south-north direction (Figure 1). The model was driven by different tidal conditions, including spring and neap tides, at the seaward boundary. The seaward boundary was set between Hartland Point in South West England and Stackpole Head in West Wales. Time varying water levels were specified along this boundary. The upstream landward boundary was set at the tidal limit of the Severn Estuary, located close to Gloucester, to account for the possible impact of the River Severn on both the residence and exposure times in the Severn Estuary. The corresponding water levels at the open boundary were specified using the predicted elevation data from POLPRED [11]. The simulation duration for calibration was 300 h, starting on 20th July and ending on 2nd August 2001.

The residence time, exposure time and the return coefficient were calculated for the interested region, i.e., the Severn Estuary (see Figure 1). Offshore surveys were carried out using the EA coastal survey vessel (csv) Water Guardian. The Water Guardian was fitted with a downward facing acoustic doppler current profiler (ADCP), which continuously monitored current velocity and direction through the water column [24]. The detailed calibration and validation were carried out in a previous study [11]. The bed roughness was the main hydrodynamic parameter used for model calibration, with the bed roughness being represented as an equivalent roughness length. The model predicted water levels were validated against the field data at Mumbles, Newport and Hinkley Point [11]. The current speeds and directions were compared to field measurements available at various sites to validate the computational accuracy of the EFDC model. The differences between the predicted and field data were calculated and the root mean squared values for the tidal levels and currents were found to be 0.2122 and 0.1857, respectively. Typical comparisons of model predicted and measured data at Southerndown and Minehead (Figure 1) are shown in Figures 2 and 3.

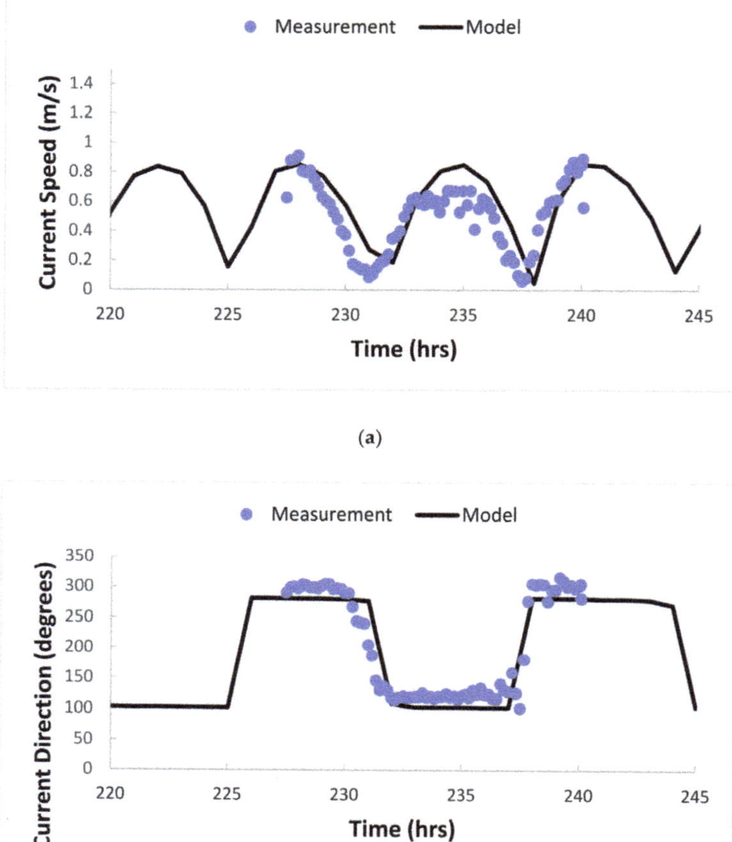

(a)

(b)

Figure 2. Comparison of (**a**) current speed and (**b**) direction at Minehead (30 July 2001).

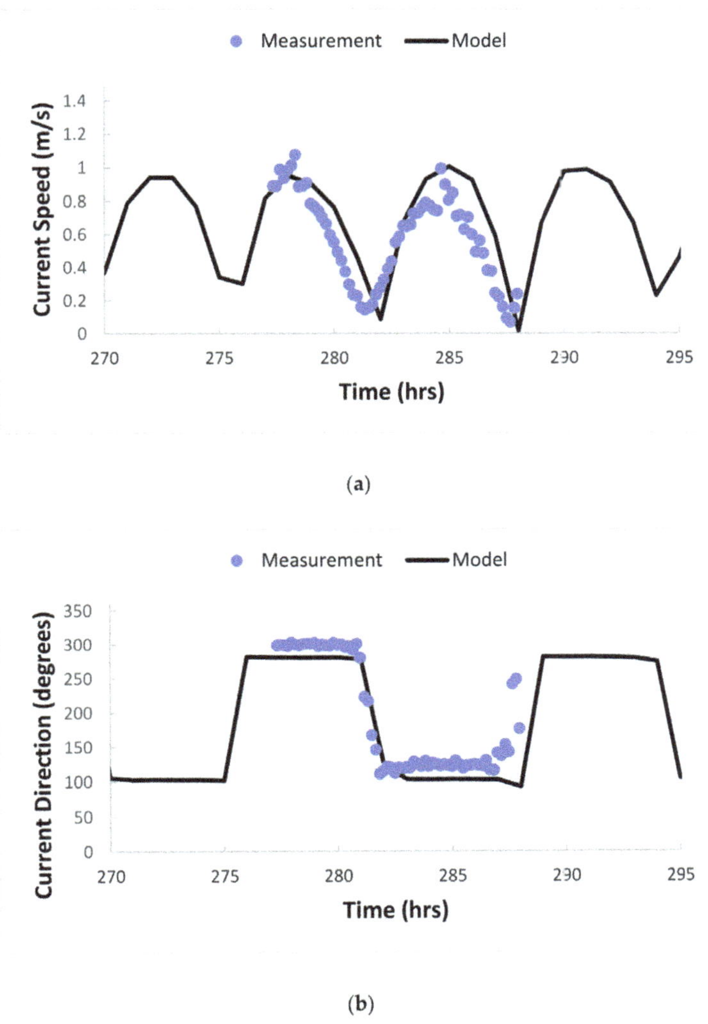

(a)

(b)

Figure 3. Comparison of (**a**) current speed and (**b**) direction at Minehead (1 August 2001).

3. Results and Discussion

In order to understand the transport time scales, i.e., the average residence time and exposure time in the Severn Estuary, 12 model scenarios were carried out for various inflows from the River Severn, tidal ranges and tracer release times. Three river inflow conditions from the River Severn were included, in the form of the base flow, average flow and high flow, and were used to represent the flow spectrum. Tracers were released at different time phases of the tide, including: SH (spring tide at high water level), SL (spring tide at low water level), NH (neap tide at high water level) and NL (neap tide at low water level).

Figure 4 shows the model simulated residence and exposure times for different river flows for the River Severn, different tidal ranges and different tracer release times in the Severn Estuary. The results (Figure 4) indicated that the exposure times were significantly higher than the residence times for all scenarios, which meant that a significant fraction of the water parcel was transported out of the Severn Estuary during ebb tides, and then returned into the basin on the subsequent flood tide for both the spring and neap tidal conditions. For neap tide conditions, the exposure time is not so much higher than the residence time as for spring tides, which indicates that for neap tide conditions a relatively smaller fraction of a water parcel, with constituents, is transported out of the estuary and returns to the estuary compared to spring tide conditions. The effects of the flow from the River Severn on the residence and exposure times were then investigated. Both the residence and exposure times decreased slightly with an increase in the river flows from the Severn, and with a decrease in the transport time scales being more significant for neap tides as compared to spring tides. Under base flow conditions, the average residence and exposure times were up to about 13.25 days and 52.87 days, respectively, while for high flow conditions, these values were reduced by 3.5% and 3.6% respectively, to 12.79 days and 50.98 days, under NL (neap tide, low water level) conditions. The numerical model predictions showed that the inflow from the River Severn under high flow conditions reduced the residence and exposure times by 1.5 to 2.4% for spring tide conditions, and 3.5 to 3.6% for neap tide conditions. The residence and exposure times were both also affected by the tracer release time. Both the residence and exposure times followed the order of: NL > NH > SL > SH, which indicated that for the Severn Estuary the transport time scale was greater for neap tide conditions, rather than spring tide conditions. However, this finding was different from the results for the macro-tidal Dublin Bay. These differences were caused by the significant variation in the return coefficient, for different tidal conditions in both studies. The differences in the return coefficients are shown in Figure 5. Here, it can be seen that the return coefficients, for both the spring and neap tide conditions, are very high, and range from 0.75–0.88. This means that, for both spring and neap tides, the basin has a strong capacity for mixing and transporting the water parcel, or tracer, out of the basin, due to the high return coefficient. This means that a large fraction of the water and tracer re-entered the estuary during the next flood tide, which was only observed during spring tides in other basins. This finding suggested that there were significant differences between micro, macro and hyper tidal basins. Therefore, this result is important for water managers responsible for maintaining high estuarine water quality, in that it is necessary to choose the most appropriate time to release any pollutants into an estuary to optimise the mixing and exchange properties and reduce the time of any pollutants in a well flushed estuary. This observation also explained the large differences between the exposure and residence times. Unlike other estuarine basins considered in this study, the neap tides of the Severn Estuary and Bristol Channel had a relatively high capacity of mixing and advection of water parcels within the basin, and a significant volume of water was flushed out of the basin on the ebb tide, and with a significant volume also re-entering the basin on the subsequent flood tide. This was not observed in a similar micro tidal estuary study [4], with the return coefficient in the micro tidal estuary showing that the coefficient was only slightly different for spring and neap tidal conditions, with typical values ranging from 0.5–0.6. In macro tidal water bodies, such as Dublin Bay, the return coefficient for neap and spring tide conditions are different, wherein for neap tides the coefficient is typically between 0.1–0.3 and with much higher values for spring tides, with typical values being in the range 0.6–0.8. For a micro tidal estuary, such as the Pearl River Estuary, the coefficients were not significantly different between neap and spring tide conditions, and with much smaller values in comparison to the hyper-tidal Severn Estuary. This suggests that different water management strategies are needed for managers responsible for designing wastewater discharge strategies into the receiving waters.

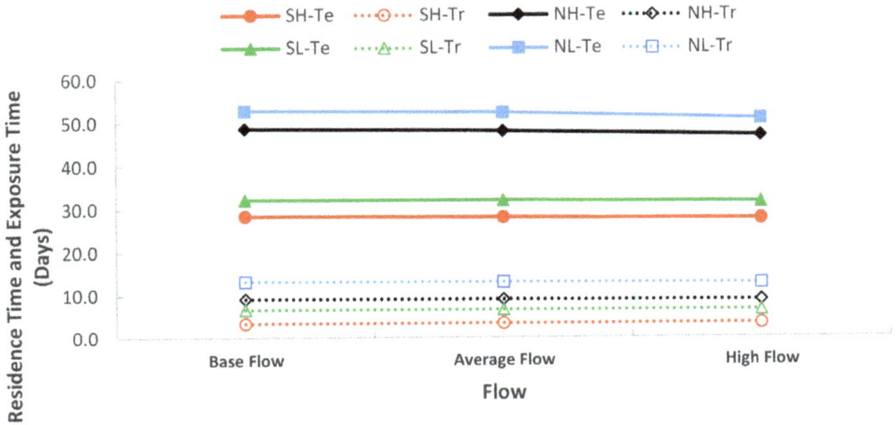

Figure 4. Variation of residence time and exposure time for different flow conditions.

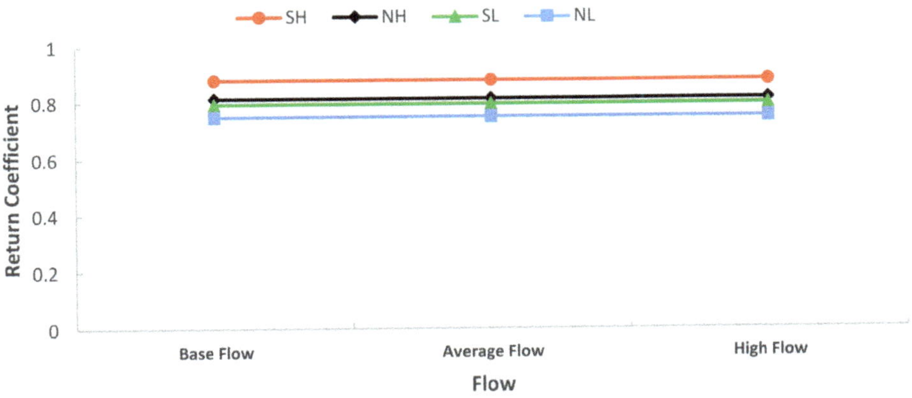

Figure 5. Variation of return coefficient.

Both the distribution of the exposure and residence times in Figure 6 confirmed that the river inflow from the Severn only affected the exposure and residence times in the upper part of the estuary under base flow conditions (Figure 6a,b,d,e), but under high flow conditions (Figure 6c,f), the effects extended further downstream. Similar patterns were observed for other modelling scenarios, including SL in Figure 7, NH in Figure 8 and NL in Figure 9. A significant difference between the exposure time (Figure 6a–c) and residence time (Figure 6d–f) was consistently observed, with higher values being predicted for the return coefficients. Figures 6 and 7 showed differences for both the residence and exposure times under SH and SL conditions. The residence time under SH was lower than for the SL condition, due to the effects of the flood tide after the initial release time, which is observed and supported by the predictions shown in Figure 4. The exposure time for the SH scenario was significantly lower than for the SL scenario (Figures 4, 6 and 7). Under SL conditions (Figure 7), the returning effects of the tide were shown only to affect noticeably the outer and deeper parts of the estuary. However, for scenario SH (Figure 6) the whole area of interest was affected. For the NH and (Figure 8) and NL conditions (Figure 9), the river inflows did not have a significant effect on the residence and exposure times, particularly in comparison with similar studies for micro and macro tidal coastal basins. The river inflow effects were more pronounced under neap tide conditions (Figures 8 and 9) than for spring tide conditions (Figures 6 and 7), but again, not as significant as observed in micro and macro tidal water basins. The returning impact for neap tides (i.e., NH, NL) were relatively small and typically in the range 0.75–0.81, with this range being typically 0.79–0.88 for spring tide conditions (i.e., SL, SH). However, both these sets of results were significantly higher than observed in other water bodies, particularly under neap tide conditions. The spatial distribution of exposure and residence times showed that there were regions of high exposure and residence times in shallow water region for low water level releases during spring (Figure 7) and neap (Figure 9) tidal conditions. The regional high transport time scale areas were not observed for high water level release of the tracers, for both spring (Figure 6) and neap (Figure 8) tidal conditions. The existence of higher transport time scale areas suggested that regional inputs of pollutants from these sites would be relatively hard to dilute efficiently through the hydrodynamic processes alone, including both river inflows and tidal processes, and if the tracer was released at lower water levels, but the overall average transport time scale was not significantly affected by the release time. The results also indicated that the transport time scale in the shallow water regions was more sensitive to the release time, which confirmed that special attention is needed by the water managers and engineers in minimising the hydro-environmental challenges in such regions.

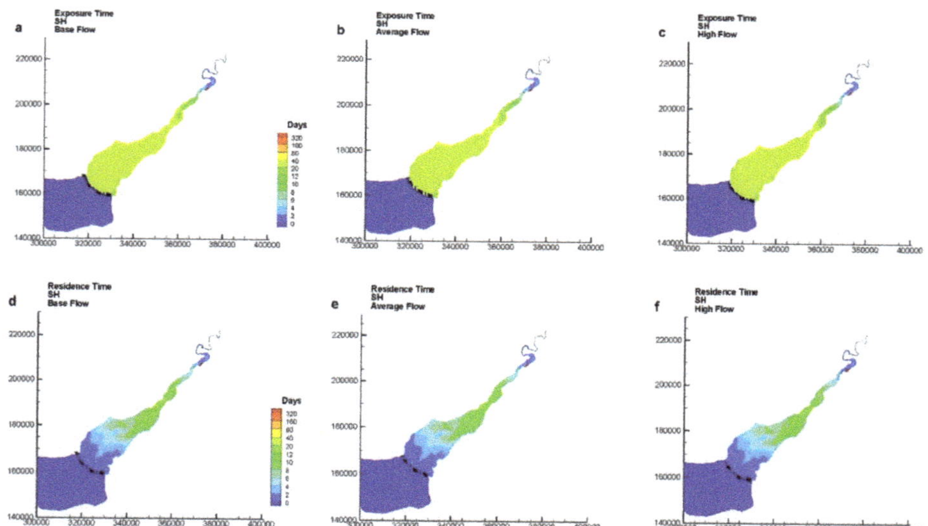

Figure 6. Exposure time (**a–c**) and residence time (**d–f**) distribution at spring tide at high water level (SH).

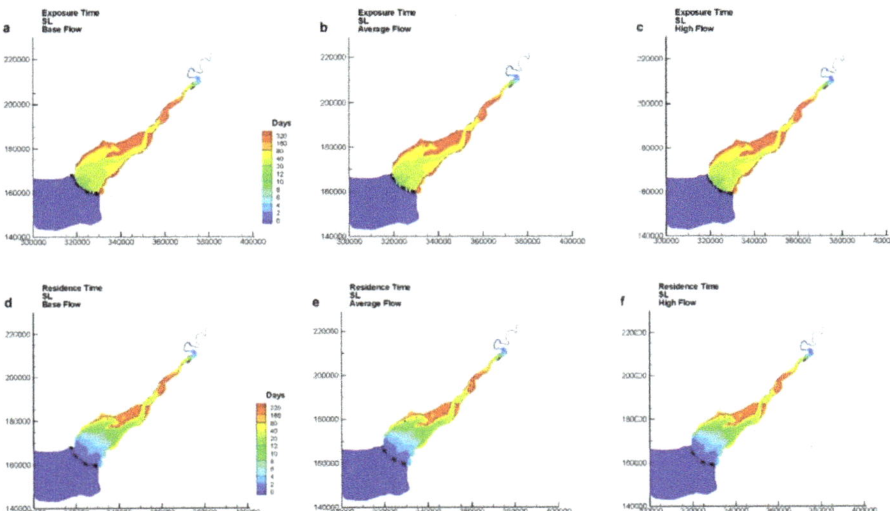

Figure 7. Exposure time (**a**–**c**) and residence time (**d**–**f**) distribution at spring tide at low water level (SL).

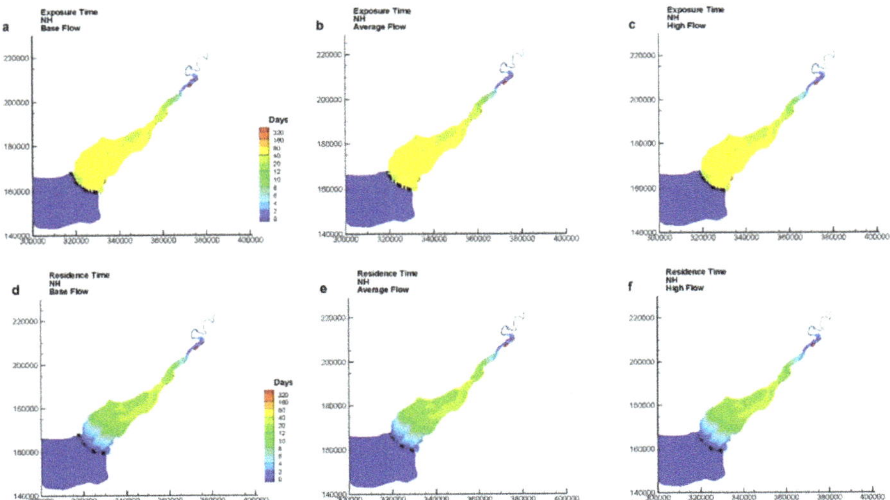

Figure 8. Exposure time (**a**–**c**) and residence time (**d**–**f**) distribution at neap tide at high water level (NH).

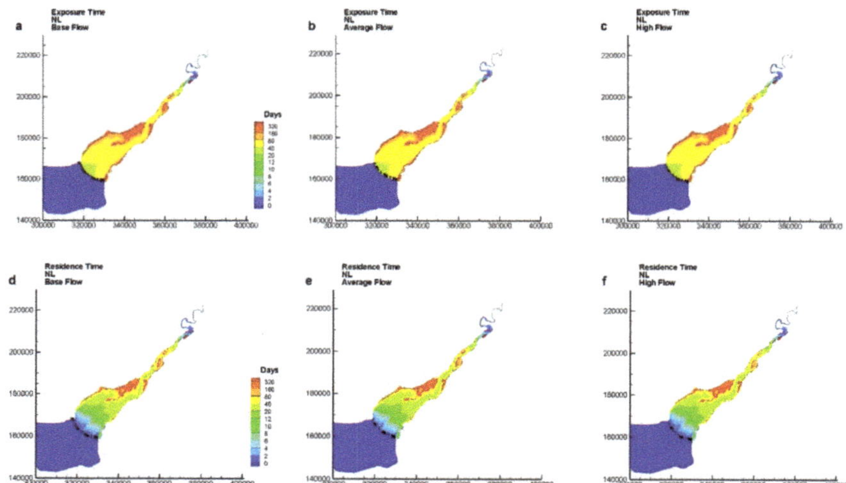

Figure 9. Exposure time (**a–c**) and residence time (**d–f**) distribution at neap tide at low water level (NL).

4. Conclusions

The main objectives of this study were to investigate the transport time scales (TTSs) of the hyper-tidal Severn Estuary by predicting and analysing the exposure and residence times. An integrated hydrodynamic and solute transport model, namely EFDC, was refined and applied to investigate the transport time scales in the Severn Estuary in the UK. Various modelling scenarios were carried out to investigate the effects of different river flow conditions, tide ranges and tracer release times on the water exchange processes for the Severn Estuary. By comparing the results obtained with macro- and micro-tidal basins, the main conclusions obtained are summarised below:

(1)   The average residence and exposure times for a hyper-tidal estuary, such as the Severn Estuary, are not significantly affected by the river flow from the River Severn. Higher river flows give only slightly smaller average residence and exposure times for all modelling scenarios, which suggests that both the exposure and residence times do not show significant seasonal variations for different river flow conditions, as compared with similar results in micro- and macro-tidal water systems.

(2)   The effects of river flows from the River Severn on the residence and exposure times in the Severn Estuary are regional in the upstream part of the estuary, for both spring and neap tidal conditions, with the effects for high flow conditions extending slightly further downstream.

(3)   The Severn Estuary is a hyper-tidal estuary with the second highest tidal range in the world, and the corresponding impact of this high tidal range on the degree of mixing and water exchange processes is, as expected, found to be significant. A previous study on micro-tidal estuaries has shown that both the exposure and residence times were lower if the tracers were released at higher water levels, regardless of the tide ranges [4]. However, the findings from this study have shown that the tidal effects in the Severn Estuary are quite different. Both the residence and exposure times followed the order of NL (neap low) > NH (neap high) > SL (spring low) > SH (spring high), which means that the tidal range plays a dominant role in the transport time scale, with the higher transport time scales being observed for neap tide conditions and particularly at low water level.

(4)   The return coefficient for the Severn Estuary does not vary significantly, with values ranging from 0.75 for the NL scenario to 0.88 for the SH scenario, while the NH scenario gave slighter higher return coefficients of 0.79 and a lower value of 0.81 for the SL scenario. The relatively high

return coefficients for both spring and neap tide conditions confirmed that there were significant differences between the exposure and residence times for all scenarios modelled.

(5) For the same tidal range conditions, releasing tracers at higher water levels gave lower residence and exposure times. For macro-tidal coastal waters, such as Dublin Bay, the effects of different return coefficients, under high tidal range conditions, meant that lower exposure times were not guaranteed, such as observed with SH > NH. However, in the hyper-tidal Severn Estuary the higher tidal ranges resulted in lower exposure and residence times. For the same tidal range, then releasing a tracer at a higher water level gave higher return coefficients in the estuary, with SH > SL and NH > NL. This result has a significant impact on designing wastewater treated discharges, particularly under extreme flood conditions.

Author Contributions: All authors have significantly contributed to the manuscript by initiating the idea (G.G., J.X.), designing the study (G.G., Y.W., J.X.), contributing to numerical modelling study for Bristol Channel and Severn Estuary (G.G., R.A.F., J.X.), and analyzing the data and results (G.G., Y.W.). G.G. drafted the manuscripts and all authors (G.G., J.X., R.A.F., Y.W.) have reviewed and agreed upon the manuscript content. All authors have read and agreed to the published version of the manuscript.

Funding: This research was funded by Open Research Fund Program of the State Key Laboratory of Water Resources and Hydropower Engineering Science, Wuhan University grant number [2017HLG02].

Acknowledgments: This work was financially supported by the Open Research Fund Program of the State Key Laboratory of Water Resources and Hydropower Engineering Science, Wuhan University (2017HLG02).

Conflicts of Interest: The authors declare that the research was conducted in the absence of any commercial or financial relationships that could be construed as a potential conflict of interest.

References

1. Gao, X.; Zhao, G.; Zhang, C.; Wang, Y. Modeling the exposure time in a tidal system: The impacts of external domain, tidal range, and inflows. *Environ. Sci. Pollut. Res. Int.* **2018**, *25*, 11128–11142. [CrossRef]

2. Monsen, N.E.; Cloern, J.E.; Lucas, L.V.; Monismith, S.G. A comment on the use of flushing time, residence time, and age as transport time scales. *Limnol. Oceanogr.* **2002**, *47*, 1545–1553. [CrossRef]

3. de Brauwere, A.; de Brye, B.; Blaise, S.; Deleersnijder, E. Residence time, exposure time and connectivity in the Scheldt Estuary. *J. Mar. Syst.* **2011**, *84*, 85–95. [CrossRef]

4. Sun, J.; Lin, B.; Li, K.; Jiang, G. A modelling study of residence time and exposure time in the Pearl River Estuary, China. *J. Hydro-Environ. Res.* **2014**, *8*, 281–291. [CrossRef]

5. Rayson, M.D.; Gross, E.S.; Hetland, R.D.; Fringer, O.B. Time scales in Galveston Bay: An unsteady estuary. *J. Geophys. Res Ocean.* **2016**, *121*, 2268–2285. [CrossRef]

6. Fischer, H.B.; List, E.J.; Koh, R.C.Y.; Imberger, J.; Brooks, N.H. *Mixing in Inland and Coastal Waters*; Academic: San Diego, CA, USA, 1979.

7. Delhez, É.J.M.; de Brye, B.; de Brauwere, A.; Deleersnijder, É. Residence time vs influence time. *J. Mar. Syst.* **2014**, *132*, 185–195. [CrossRef]

8. Shang, J.; Sun, J.; Tao, L.; Li, Y.; Nie, Z.; Liu, H.; Chen, R.; Yuan, D. Combined effect of tides and wind on water exchange in a semi-enclosed Shallow Sea. *Water* **2019**, *11*, 1762. [CrossRef]

9. Gao, X.; Chen, Y.; Zhang, C. Water renewal timescales in an ecological reconstructed lagoon in China. *J. Hydroinform.* **2013**, *15*, 991–1001. [CrossRef]

10. Lyddon, C.; Brown, J.M.; Leonardi, N.; Plater, A.J. Flood hazard assessment for a hyper-tidal estuary as a function of tide-surge-morphology interaction. *Estuaries Coasts* **2018**, *41*, 1565–1586. [CrossRef]

11. Zhou, J.; Falconer, R.A.; Lin, B. Refinements to the EFDC model for predicting the hydro-environmental impacts of a barrage across the Severn Estuary. *Renew. Energy* **2014**, *62*, 490–505. [CrossRef]

12. Gao, G.; Falconer, R.A.; Lin, B. Modelling importance of sediment effects on fate and transport of enterococci in the Severn Estuary, UK. *Mar. Pollut. Bull.* **2013**, *67*, 45–54. [CrossRef] [PubMed]

13. Uncles, R.J. Physical properties and processes in the Bristol Channel and Severn Estuary. *Mar. Pollut. Bull.* **2010**, *61*, 5–20. [CrossRef] [PubMed]

14. Uncles, R.J.; Stephensa, J.A.; Smithb, R.E. The dependence of estuarine turbidity on tidal intrusion length, tidal range and residence time. *Cont. Shelf Res.* **2002**, *22*, 1835–1856. [CrossRef]

15. Viero, D.P.; Defina, A. Water age, exposure time, and local flushing time in semi-enclosed, tidal basins with negligible freshwater inflow. *J. Mar. Syst.* **2016**, *156*, 16–29. [CrossRef]

16. Zimmerman, J.T.F. Mixing and flushing of tidal embayments in the Western Dutch Wadden Sea, Part I: Distribution of salinity and calculation of mixing timescales. *Neth. J. Sea Res.* **1976**, *10*, 149–191. [CrossRef]

17. Andutta, F.P.; Helfer, F.; de Miranda, L.B.; Deleersnijder, E.; Thomas, C.; Lemckert, C. An assessment of transport timescales and return coefficient in adjacent tropical estuaries. *Cont. Shelf Res.* **2016**, *124*, 49–62. [CrossRef]

18. Delhez, É.J.M. On the concept of exposure time. *Cont. Shelf Res.* **2013**, *71*, 27–36. [CrossRef]

19. Delhez, É.J.M.; Heemink, A.W.; Deleersnijder, É. Residence time in a semi-enclosed domain from the solution of an adjoint problem. *Estuar. Coast. Shelf Sci.* **2004**, *61*, 691–702. [CrossRef]

20. Hamrick, J.M. *A Three-Dimensional Environmental Fluid Dynamics Computer Code: Theoretical and Computational Aspect*; Hamrick, J.M., Ed.; Special Report in Applied Marine Science and Ocean Engineering; No. 317; Virginia Institute of Marine Science, School of Marine Science, The College of William and Mary: Gloucester Point, VA, USA, 1992; p. 63.

21. Du, J.; Shen, J. Water residence time in Chesapeake Bay for 1980–2012. *J. Mar. Syst.* **2016**, *164*, 101–111. [CrossRef]

22. Gao, X.; Xu, L.; Zhang, C. Estimating renewal timescales with residence time and connectivity in an urban man-made lake in China. *Environ. Sci. Pollut. Res. Int.* **2016**, *23*, 13973–13983. [CrossRef]

23. Xia, J.; Falconer, R.A.; Lin, B. Hydrodynamic impact of a tidal barrage in the Severn Estuary, UK. *Renew. Energy* **2010**, *35*, 1455–1468. [CrossRef]

24. Stapleton, C.M.; Wyer, M.D.; Kay, D.; Bradford, M.; Humphrey, N.; Wilkinson, J.; Lin, B.; Yang, Y.; Falconer, R.A.; Watkins, J.; et al. *Fate and Transport of Particles in Estuaries*; Environment Agency Science Report; Environmental Agency: Bristol, UK, 2007; SC000002/SR1-4.

Article

Observation Strategies Based on Singular Value Decomposition for Ocean Analysis and Forecast

Maria Fattorini [1,2] **and Carlo Brandini** [1,2,*]

1   Institute of BioEconomy (IBE, C.N.R.), via Madonna del Piano 10, 50019 Sesto Fiorentino, Italy; maria.fattorini@ibe.cnr.it

2   LaMMA Consortium—Environmental Modelling and Monitoring Laboratory for Sustainable Development, via Madonna del Piano 10, 50019 Sesto Fiorentino, Italy

*   Correspondence: brandini@lamma.toscana.it; Tel.: +39-(055)-448-3052

Received: 31 October 2020; Accepted: 3 December 2020; Published: 8 December 2020

Abstract: In this article, we discuss possible observing strategies for a simplified ocean model (Double Gyre (DG)), used as a preliminary tool to understand the observation needs for real analysis and forecasting systems. Observations are indeed fundamental to improve the quality of forecasts when data assimilation techniques are employed to obtain reliable analysis results. In addition, observation networks, particularly in situ observations, are expensive and require careful positioning of instruments. A possible strategy to locate observations is based on Singular Value Decomposition (SVD). SVD has many advantages when a variational assimilation method such as the 4D-Var is available, with its computation being dependent on the tangent linear and adjoint models. SVD is adopted as a method to identify areas where maximum error growth occurs and assimilating observations can give particular advantages. However, an SVD-based observation positioning strategy may not be optimal; thus, we introduce other criteria based on the correlation between points, as the information observed on neighboring locations can be redundant. These criteria are easily replicable in practical applications, as they require rather standard studies to obtain prior information.

Keywords: singular value decomposition; data assimilation; ocean models; observation strategies; ocean forecasting systems; ocean Double Gyre; 4D-Var; ROMS

---

1. Introduction

In recent years, there has been a growing demand for oceanographic forecast data [1,2], which comes from different public and private subjects for operational oceanography purposes. This request stimulates the production of reliable predictions of physical and biogeochemical ocean variables to support activities such as search and rescue operations, ocean energy, fisheries, and environmental monitoring and pollution control. Observations play an essential role in operational systems as they also allow evaluating the reliability of predictions. The most important initiative in this context is the Global Ocean Observation System (GOOS), which includes several regional observation components providing data to global and basin-scale operational services, as well as to regional downstream services [3].

Operational oceanographic services both at the basin and regional scales improve their forecast reliability when the model forecast is properly initialized with fields obtained through a data assimilation procedure. Data assimilation (DA) combines observations and models first-guess-weighted by their respective accuracies to obtain the best unbiased estimation of the ocean state. In a DA scheme, the observations correct the trajectory (first guess) according to their influence, which mainly depends on the observation and model error covariance matrices. As a consequence, DA can be useful to better control the error growth of the state trajectory with respect to the real evolution. Furthermore, in the operational practice, a common procedure of initializing a simulation starting with external data (e.g., climatology, objective analysis, model analysis, etc.) requires a spin-up interval during which the solution is not usable. DA schemes as 4D-Var reduce the spin-up effects (keeping a dynamical consistency between analysis and model equations) and also reduce model uncertainties.

Large amounts of data come from satellite observations (mainly Sea Surface Temperature and Sea Surface Height), which have some intrinsic limitations (surface-limited observations, revisiting times). Many parameters are only observable by collecting in situ observations through specific sensor networks that integrate satellite observations with data along the water column and at higher frequencies. The main limitation of in situ observation networks is their high cost for installation and maintenance over time; it is very important, therefore, to design an in situ observing system that maximizes the impact of the observations in the forecast, minimizing the cost.

Ocean models can also be used to evaluate both existing and new observing networks through different methodologies [4]. Observing System Experiments (OSEs) compare analysis obtained by eliminating only a part of the observations with the analysis obtained by assimilating the entire dataset to understand the impact of the omitted observations. Observation System Simulation Experiments (OSSEs) use "synthetic" observations to evaluate the benefit of assimilating observations from instruments/networks not yet installed. Adjoint-based techniques and ensemble-based methods can be used to study observation sensitivities and the impact on assimilated systems, contributing to the design of observing systems [4–6].

As different observations have different impacts when they are assimilated in an ocean model, a major problem is designing an observation network that provides data giving the best results (i.e., fewer errors) when they are assimilated. The positioning of the observing system is indeed somehow related to the unstable modes, which deserve more than others to be corrected. Since the fundamental milestone made by Lorenz in 1965 [7], it is well known how the divergence in chaotic systems rises from the unstable directions of the state trajectory where small errors in initial conditions significantly grow, leading to very different final states. This places a time limit to the predictability of the system state, which is usually evaluated by the largest Lyapunov exponents. The assimilation of observations attempts to prolong that time limit [8]. Some significant errors could decay over time, whereas smaller errors could intensely grow and produce a heavy impact on forecast reliability. The growth of the divergence between model evolution and the real state of the system is driven by these unstable directions, rather than by the largest components of the error embedded in the predicting system [9,10]. Indeed, the structure of the fast-growing perturbations is a dynamically evolving field and depends on the flow regime, as it derives from the position of the current state on the attractor and varies over time [7,11,12].

For what concerns observation strategies, we can expect that a suitable positioning of observation devices in areas in which error in the initial condition is fast-growing may better control this growth. Such a choice can be performed on the basis of the study of perturbation growth. Hansen and Smith [13] showed that for sufficiently small errors, observation strategies based on system dynamics produce better results than strategies based on error estimates.

A notable contribution to this field was made by Farrell and Joannou [14,15] in their General Stability Theory (GST) of a dynamical system, in which they extended the traditional modal stability theory to include transient growth processes. The authors identified the decomposition to singular values (SVD) as a suitable tool for treating perturbation growths in geophysical fluid systems. A variation of this method considers the calculation of Hessian singular vectors, which identify the errors in initial conditions that evolve into the largest forecast errors [6].

The existence of large singular values indicates that small errors in the initial background state can grow very rapidly, reducing the system predictability, and the respective singular vectors indicate the areas where disturbances grow faster. Analyzing Singular Vectors (SVs) appears strategic to increase model predictability by giving an indication of where it is more important to reduce errors in initial conditions [6,16].

The application of SVD to select observations has been already tested in a number of studies, mostly related to operational aspects of atmospheric forecasting systems [5,17–21]. A review of observation strategies [22] confirmed the utility of SVD information in choosing the observations to be assimilated. Bergot et al. [23] considered the SVD as a useful tool to identify areas where assimilating even a few observations is able to significantly reduce forecast errors. An important portion of literature about the topic of adaptive observations was originated by the first experimental reference, the so-called Fronts and Atlantic Storm Track Experiment (FASTEX) [24–26]. Other experiments were carried out by assimilating additional observations from aircraft in regions characterized by rapidly growing SVs [19,27,28].

In this work, we test some possible observation strategies of a simplified ocean system with the aim of establishing an optimal configuration of an in situ observing network able to reduce forecasting uncertainties through DA. For simplicity, we limit this study to velocity observations. In particular, our goal is to achieve such an optimal configuration using a limited number of observation points, as these may have a significant cost, in order to ensure the greatest possible benefit to an integrated assimilation/forecasting system. The benefit is measured with respect to the short-/medium-range forecast (analysis and forecast cycles of a few days), as required in the operational practice. In order to select the possible observation points, the proposed strategy is based on the SVD and on a maximum correlation among the horizontal velocities, which is traduced into a minimum distance, variable over the model domain. Indeed, we verify that an observation strategy based on SVD may fail if it is not accompanied by other considerations linked to the flow structure, and such a combination of SVD analysis with a correlation analysis can be used to limit redundant observations.

The experiments are carried out by using the ocean model ROMS (Regional Ocean Modelling System, www.myroms.org) [29], which already includes suitable routines for the SVD computation [30]. We perform a set of numerical experiments assimilating different datasets and investigate the effects on model results.

In Section 2, the configuration of the experiment is presented, and in particular, the description of the model, the DA scheme, and the proposed strategy to place in situ observation points.

In Section 3, the results of all experiments are reported; our best strategy is compared to a random localization strategy and also to a selection procedure based on SVD and on a minimum fixed distance among observations.

2. Materials and Methods

2.1. The Model Set-Up

The reference model used in this work is an idealized ocean model, widely known as Double Gyre (DG). Although it is conceptually much simpler than realistic simulations, it is still relevant from an oceanographic point of view. DG simplified dynamics have been used as an idealized case to reproduce the seasonal and interannual oscillations of the large-scale circulation in the ocean, useful for the climate system predictability [30].

This configuration is also used by the ROMS developers [31] as a test case to introduce the functionalities of the tangent linear model and its adjoint. This simplified ocean system is strongly barotropic, and no relevant differences can be found in different vertical layers. The basin has a flat bathymetry 500 m deep with all closed boundaries. The model domain is a large rectangular basin 1000 km in size in the east–west direction and 2000 km in the north–south direction; it is discretized horizontally in 56 × 110 cells and vertically in 4 equally spaced s-levels. The model is forced by a constant zonally uniform wind stress with a positive zonal component at its midaxis, which is inverted to a negative zonal component approaching the northern and southern boundaries, as defined by a sinusoidal function of latitude:

$$\tau_x = -\tau_0 cos\left(\frac{2\pi y}{L_y}\right) \tag{1}$$

where $\tau_0 = 0.05$ N/m^2 and L_y is the meridional extent of the basin. This particular wind distribution induces two large interacting vortices with a scale of about 1000 km: a subpolar cyclonic gyre and a subtropical anticyclonic gyre, whose stationary depends on the eddy viscosity values.

The model solves the 3D primitive equations in a beta-plane approximation centered at 45° N. Density profiles are defined by an analytically stable profile of the active tracers, as described by Moore et al. [31]. The advection for both 3D momentum and tracers in the horizontal and vertical components is implemented by, respectively, the third-order upstream scheme along constant S-surfaces and the fourth-order centered scheme. Horizontal turbulent processes are parametrized by a harmonic operator whose horizontal eddy viscosity and diffusivity are equal to 160 m^2/s. Although forced by a steady wind, the model solution depends on the eddy viscosity value, and circulation passes from a stationary pattern (high eddy viscosity) to an oscillating behavior (low eddy viscosity), with the formation of smaller-scale vortices and the shifting of the main current patterns. In fact, the DG circulation shows a bifurcation [32], corresponding to a critical value of the Reynolds number. For values lower than the critical value, the flow converges quickly to a unique steady solution, whereas for values above the critical value, instabilities occur, and the symmetry of the structure of the subpolar and subtropical cells is broken. The vertical turbulent mixing is parameterized by coefficients of vertical viscosity and diffusion of 1 m^2/s; linear bottom friction is introduced with a bottom drag coefficient of 8×10^{-7} m/s. To ensure model stability, the numerical time steps are set, respectively, to 45 s (barotropic time step) and 900 s (baroclinic time step).

The model is initially started through the steady solution (Figure 1a), obtained by a 20-year-long simulation with a high eddy viscosity, equal to 1280 m^2/s, which ensures a steady circulation that is symmetrical with respect to the zonal axis. For the following 10-year run, the eddy viscosity is then decreased to a lower value of 160 m^2/s: in this case, the circulation loses symmetry as it becomes unsteady and characterized by meandering where the two original gyres move through the domain with other gyres arising.

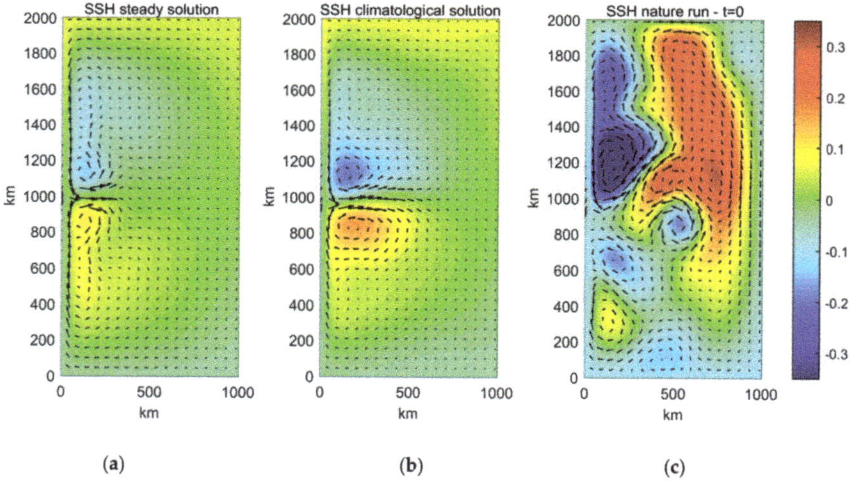

Figure 1. Surface current field (arrows) and Sea Surface Height (SSH) (m) of (**a**) the steady solution, (**b**) the climatological state, and (**c**) the initial state of the Nature Run for the first assimilation window.

For our model experiments we define a couple of independent model runs:

1. A Nature Run (NR), which is the model initialized by the last snapshot of the 10-year run (Figure 1c). This is considered the true state of the ocean, a virtual reality from which synthetic observations are extracted;

2. A Free Run (FR) or background, which is the simulation initialized by an initial velocity field obtained by averaging the unsteady solution of the 10-year run (Figure 1b). This represents the common and simplest way to initialize numerical simulations by means of a "climatological" time-averaged solution.

In each experiment, a set of observations is extracted from the NR using the observation strategies discussed below. Such a synthetic dataset includes velocity observations in the form of ocean current profiles at a frequency of 15 min, which are assimilated using the FR as the background, thus producing an analysis. In real ocean-observing systems, such kinds of observations are normally taken by means of Acoustic Doppler Current Profilers (ADCPs). The analysis is performed in a 5-day assimilation window that provides initial conditions to a subsequent 5-day forecast.

The positions of observation instruments are fixed in each experiment and chosen by adopting three different strategies: (1) randomly, (2) in the area of maximum dominant singular vectors and imposing a minimum distance among the positions, and (3) in the area of maximum dominant singular vectors, imposing a maximum limit of correlation between the velocity time series in each position. For what concerns the first set of experiments (random positioning), the positions are obtained by the rand function in MATLAB©, assuming a minimum distance is equal to the grid cell size (in this case, around 18 km). To reduce the dependence of the results on a particular random spatial configuration, we repeat each experiment by assimilating different datasets characterized by the same number of (virtual) observation instruments. Finally, to assess the selection procedure for different hydrodynamic states, the assimilation test is repeated in different time windows.

The assimilation of synthetic observations is executed through the incremental formulation of the 4D-Var based on the Lanczos algorithm (ROMS-IS4DVar). In such a procedure, the increments to add to the control vector to minimize the cost function are computed iteratively [5]. The control vector corresponds to the initial state, so only the initial conditions were adjusted by assimilating data.

A maximum number of 10 inner loops and 2 outer loops are set up, and the assimilation process stops when the minimization of the cost function gradient reaches a given tolerance that we set equals 10^{-4}, as we verified that this limit is sufficiently adequate for convergence. The presence of observation errors is considered in the observation error matrix as we assume implicitly that observations are affected by an error of 0.01 m/s.

The background error covariance matrix B_x is factorized by means of the univariate correlation matrix C, the diagonal matrix of the prior error standard deviations Σ_x, and the multivariate balance operator K_b, as described in [5]:

$$B_x = K_b \Sigma_x C \Sigma_x^T K_b^T. \tag{2}$$

The background error standard deviation Σ_x is defined by using the standard deviation of the state variable fields during the 10-year-long (unsteady) simulation, a period long enough to compute meaningful circulation statistics.

The correlation matrix C is in turn factorized by a diagonal matrix of grid box volumes W, a horizontal and vertical correlation function model L_h and L_v, and a matrix of normalization coefficients Λ, as:

$$C = \Lambda L_v^{1/2} L_h^{1/2} W^{-1} L_h^{T/2} L_v^{T/2} \Lambda^T. \tag{3}$$

The normalization coefficients are used to ensure that the diagonal elements of the associated correlation matrix C are equal to unity and they are computed through the so-called "exact method," in which the horizontal and vertical isotropic decorrelation scales imposed equal 30 km and 100 m, respectively, for all the state variables.

2.2. Observation Strategies

The proposed criteria for identifying the most suitable positions to install observation instruments join two requirements: (a) finding areas characterized by high values of the dominant SVs; (b) imposing a maximum limit to the correlation between the velocity time series at the observation points.

For what concerns the first requirement, we resort to the Singular Value Decomposition (SVD) of the tangent propagator to identify the directions of maximum error growth in a given time interval, according to the Generalized Stability Theory (GST) by Farrell and Ioannou [14,15] for nonautonomous systems. SVD decomposes the matrix L (in this study, it is the tangent propagator) in three matrices that satisfy LV = US [33]. The matrices V and U are formed by the eigenvectors of L^TL and LL^T, respectively, which are autonomous and symmetric for construction. Therefore, their eigenvectors form two orthogonal bases of the domain space and the range space, respectively. For this property, any disturbance at the initial time can be written as a linear combination of the initial singular vectors in the domain space, whereas its evolution can be written as a linear combination of the final singular vectors in the range space. Furthermore, the eigenvalues are the same for both matrices, L^TL and LL^T, and are called singular values; their square root corresponds to the growth factor of the relative initial singular vector, as they are transformed by L. Therefore, the growth rate of all the perturbations is confined by the fastest-growing singular mode, characterized the higher singular value. The dominant initial singular vector defines the direction of maximum growth error in the interval $[t_1,t_2]$. This calculation allows us to assess the fastest-growing disturbance among all those possible during a given finite time interval, called optimization time. The SVD computation is controlled by two main parameters: the norm, by which the growth of the singular vectors is evaluated, and the optimization time. The dominant initial singular vector is calculated as the singular vector that maximizes a norm at the final time of the interval, as adopting different norms leads to different SVs. In most oceanographic studies, the norm used is the total energy, the same we apply in this work, although other norms have been used depending on the particular aim, such as kinetic energy, enstrophy, and the Hessian norm [5,6,17,33,34]. More details on the SVD can be found in Farrell and Ioannou [14,15].

The number of singular modes of the system is of the order of 10^5, equivalent to the dimension of the model state. The leading singular vectors are computed in ROMS through the Lanczos algorithm [35] by integrating the tangent linear model forward in time and the adjoint model backward in time a sufficient number of times for the convergence of the algorithm [31].

The SVD is applied with respect to the tangent model of the free-run circulation starting from the climatological state so it is the same for all the time windows. The 200 dominant SVs are computed for different optimization times (5, 10, 20, and 60 days). In Figure 2, the singular values computed with optimization times of 5 and 60 days are reported as examples, and we see that the choice of limiting the calculation to the first 200 SVs, containing the main information on the perturbation evolution, is adequate as it also avoids the computation of a huge number of SVs.

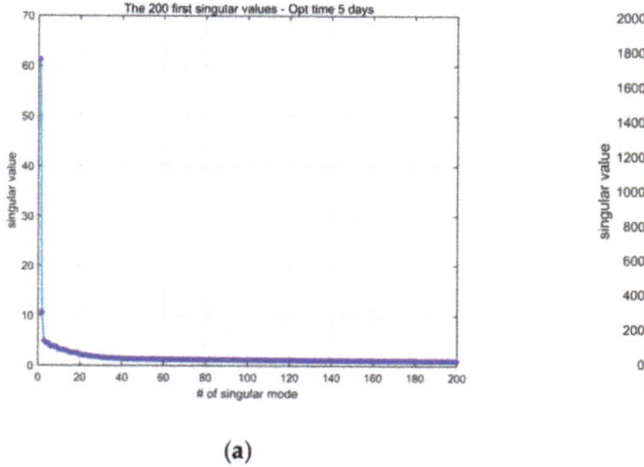

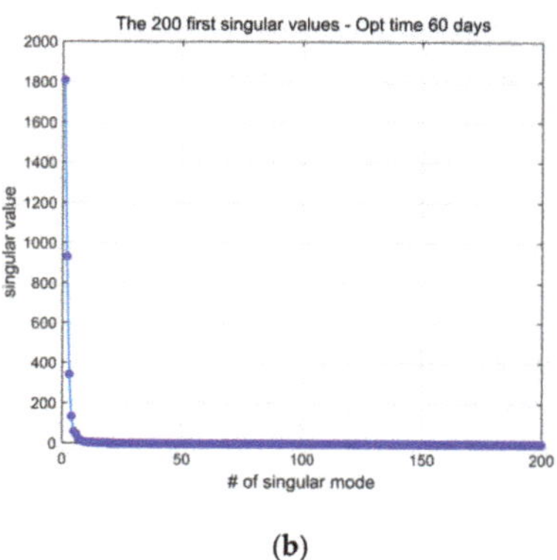

(a) (b)

Figure 2. The 200 dominant singular values computed on an optimization time of 5 days (**a**) and an optimization time of 60 days (**b**).

Figure 3 shows the sum of projections onto the horizontal surface velocity components of all the dominant initial singular vectors (upper maps) and the dominant final vectors (bottom maps), weighted with the corresponding singular values. SVs and singular values are computed from the free run by considering different optimization time (T_{op}) of 5, 10, 20, and 60 days. The structures of the initial and final singular vectors computed on a short T_{op} are similar to each other and concentrated near the convergence of the currents along the western side, where the two branches of the currents merge and change their direction (from meridional to zonal).

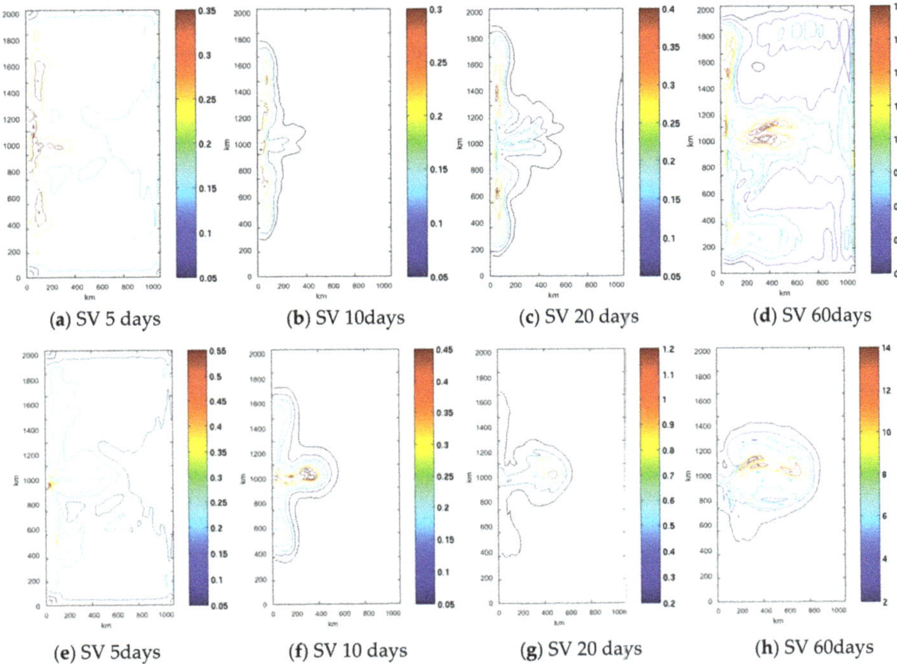

(a) SV 5 days **(b)** SV 10days **(c)** SV 20 days **(d)** SV 60days

(e) SV 5days **(f)** SV 10 days **(g)** SV 20 days **(h)** SV 60days

Figure 3. Sum of projections on the horizontal surface velocities of the 200 dominant initial singular vectors, for different optimization times (**a**–**d**). Similar sum of projections computed for the first 200 final singular vectors (**e**–**h**). Singular values are used to weight Singular Vectors (SVs).

The difference between the initial and the final singular vectors maps grows for increasing optimization times, as well as the singular values (Figure 2). The perturbations computed for 60 days evolve much more than those computed over a shorter period, in agreement with the higher singular values obtained for larger optimization times. However, significantly large singular values, even for the 5-day optimization window, are due to the fact that the initial period of the background run is subject to a strong tendency to change the circulation structure, especially close to the convergence area. Indeed, the simulation starts from the climatological state (Figure 1b) and the circulation is subject to strong variations which are concentrated in the central–western area.

Concerning the second requirement, we set a minimum distance among observations, defined on the basis of a maximum correlation between the time series of model velocities at the observation positions themselves.

The need to impose a minimum distance among in situ observations arises from the typical structure of the dominant singular vectors quite concentrated in specific areas. The highest values of the SV component on the surface velocity are strictly localized in the middle of the western boundary where the ascending and descending currents converge (Figure 2). On the one hand, it is crucial to control such areas of maximum error growth with sufficient detail. On the other hand, having too many observation points concentrated in this part would not give a strong benefit to the analysis since horizontal velocities in this area can be strongly correlated to each other. Therefore, a procedure only based on SV would lead to select too dense instrument positioning. Hence, some experiments are realized imposing a fixed minimum distance between the observation points, but this method does not avoid choosing sampling points excessively correlated. Moreover, we expect for most hydrodynamic fields of interest, the correlation between points is not homogeneous on the whole domain, as well as the suitable minimum distance among observations. This correlation is computed by the following steps:

(1) For each grid cell (i, j) we compute the spatial correlation of both the u and v velocity components with respect to all of the other grid cells (h, l), with $i \neq h$ and $j \neq l$, and then we take the norm of the matrix:

$$\begin{vmatrix} \langle u_{ij}u_{hl} \rangle & \langle v_{ij}u_{hl} \rangle \\ \langle u_{ij}v_{hl} \rangle & \langle v_{ij}v_{hl} \rangle \end{vmatrix}. \tag{4}$$

Each cell is associated with a single normalized correlation map, whose values range between 0 (uncorrelated grid cells) and 1 (perfectly correlated grid cells).

(2) A maximum value for such normalized correlation is imposed as a threshold to calculate for each grid cell the averaged distance beyond which the correlation is lower than the chosen threshold. In this way, all correlation maps for each grid cell can be transformed into a single map of distances. This map of minimum distances among in situ observation points is computed as the mean radius inside which the correlation of the variable of interest is higher than the imposed correlation threshold.

Figure 4 shows the map of distances used in this study (in km), computed by imposing a maximum correlation of the time series of velocities equal to 0.6. Note that distances are never too small (in this case, >100 km) and they increase moving away from the critical area of convergence.

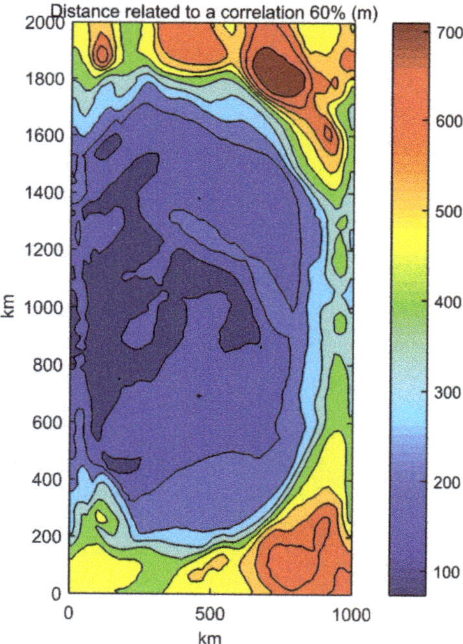

Figure 4. Map of the correlation distances (km) related to a correlation threshold of 0.6.

3. Results

As described in the previous sections, we test three different observation strategies applied to an idealized ocean model (DG) to identify the observation network configuration, which gives rise to the best analysis and forecast.

To compare each test, we adopt the Taylor diagram representation [36], which is often used in the operational field and allows collecting the same graph three of the most used statistical indices: the correlation, the centered root-mean-square error, and the standard deviations between two series (considering the Nature Run as the target). As in this study, we compare the maps of values at the same time, and correlation must be understood as a spatial correlation. These statistics are computed on the surface velocity components at the final time of the assimilation window, which is the initial condition for the five-day forecast run.

In the first experiment, we test the assimilation of an increasing number of velocity profiles, randomly located using the criteria described in the previous section. We start with 20 observation instruments (i.e., velocity profilers or ADCPs). As randomness can produce datasets more or less impactful for DA, the test is repeated considering different positions.

For the five-day assimilation window, two points are highlighted in Figure 5:

1. The Nature Run (NR), that we assume as the true state of the system;
2. The Free Run (FR), starting from the climatological state, which is poorly correlated with the NR (around 50%).

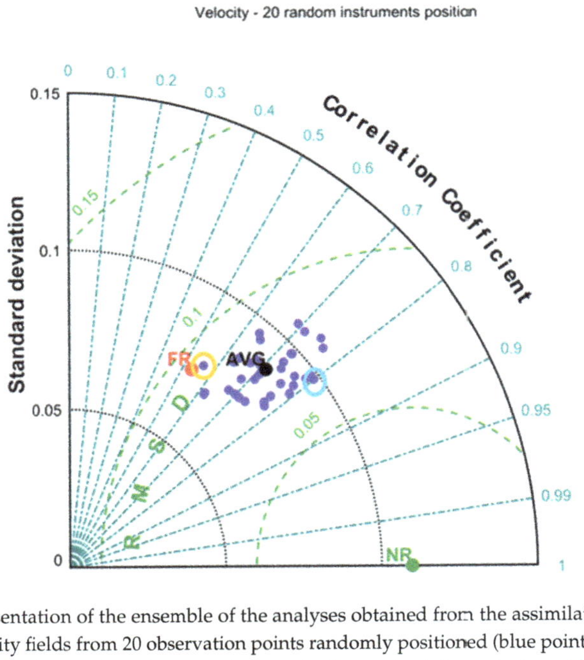

Figure 5. Representation of the ensemble of the analyses obtained from the assimilation of different datasets of velocity fields from 20 observation points randomly positioned (blue points), the Free Run (red point), and the Nature Run (green point) in the first assimilation window. The orange circle indicates the M dataset, whereas the cyan circle indicates the U dataset, which Figure 6 refers to.

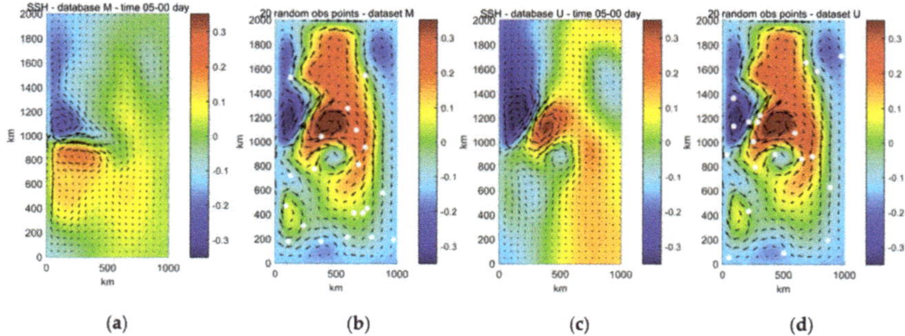

Figure 6. Vector maps of the analysis from the configuration network "M" (**a**) and "U" (**c**) at the end of the assimilation window (see Figure 5). The observation positions, taken from the Nature Run (NR), are represented in (**b**) (Dataset "M") and (**d**) Dataset "U").

We find an excellent capability of the DA algorithm (ROMS-IS4DVar) to adjust the initial condition and bring the evolution of the state of the system closer to the truth.

Figure 5 shows a wide spread between the analyses produced by assimilating different datasets, each corresponding to a different network configuration. We observe a significant spread of the results around such average, as analysis data can have a stronger (around 0.8 for Dataset M) or weaker correlation (around 0.6 for Dataset U). Some analyses although characterized by a high correlation to the NR have lower standard deviations than the NR, hence they poorly represent the true circulation structures.

As an example, in Figure 6, we report the snapshots of the circulation of both the worst and the best analysis obtained, respectively, from the synthetic datasets corresponding to the configuration networks "M" and "U."

This first set of experiments is extended through the assimilation of a growing number of randomly positioned ADCPs (40, 60, 80, 100, 150, and 200).

Looking at the averaged statistics of analyses (named AVG in Figure 7), we have a positive effect of DA in improving the estimation of the model state as the number of observation points increases. At the same time, the statistical indices within each ensemble are increasingly closer to each other.

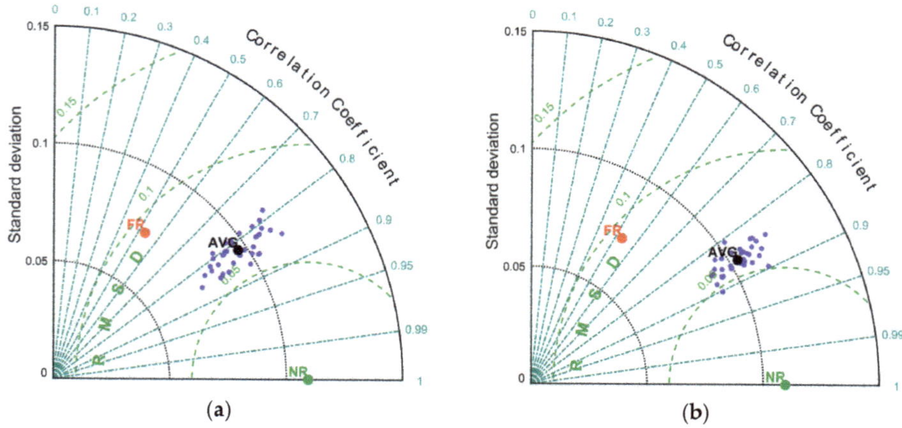

Figure 7. *Cont.*

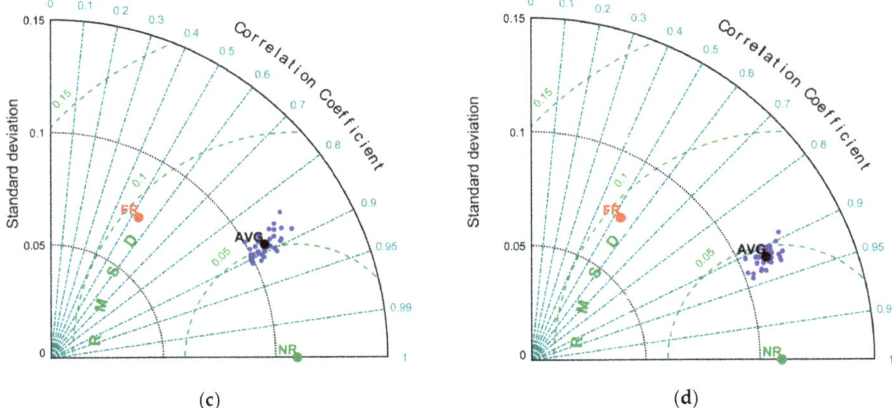

(c)

(d)

Figure 7. Representation of the ensemble of the analyses obtained from datasets of velocity measurements from, respectively, 40 (**a**), 60 (**b**), 80 (**c**), and 150 (**d**) observation points randomly positioned (blue points), Free Run (red point), and Nature Run (green point) in the first assimilation window.

The results of all tests with the same number of observation points are summarized by means of a number of average points, each representing the averaged statistics of the related ensemble in Figure 8. We observe a progressively better quality of analyses in terms of all statistical indices (correlation, RMSE, and standard deviation) and a progressively lower benefit in the assimilation of additional observations. Indeed, as the number of observations increases, the marginal improvement of the analyses decreases. Therefore, a suitable strategy to localize measuring tools is especially impactful in observation networks characterized by a few instruments, which is the case of most in situ observation networks used in operational oceanography.

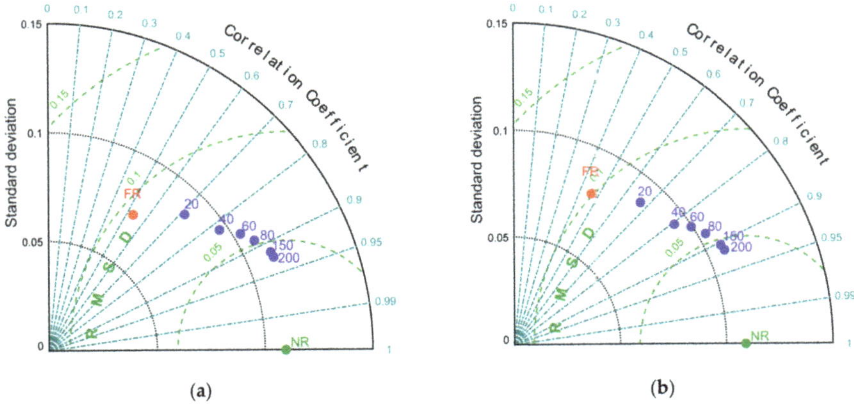

(a)

(b)

Figure 8. Representation of the average point representing a different ensemble of (**a**) the analyses obtained from datasets of velocity measures characterized by a different number of observation points randomly positioned: 20, 40, 60, 80, 150, and 200; (**b**) the subsequent forecast.

Figure 9 shows the improvement of the quality of analysis and forecast for an increasing number of observations. Forecast reliability, in terms of correlation and error, is strictly dependent on the quality of initial conditions; therefore, it is quite proportioned to the analysis.

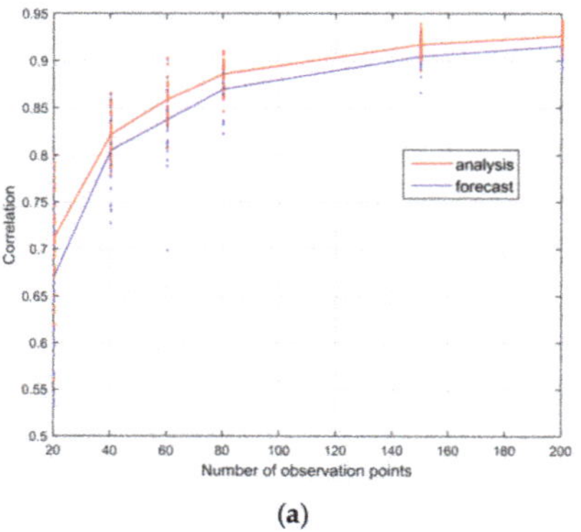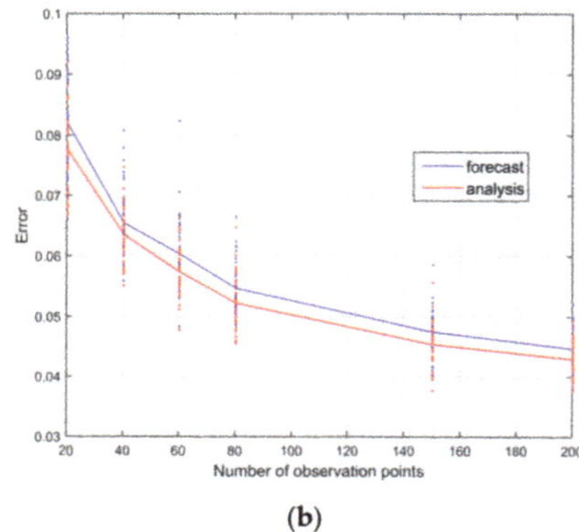

(a)   (b)

Figure 9. Correlation (**a**) and error (**b**) of the average point representing a different ensemble size (whose members are reported by single dots) of the analyses and the subsequent forecast obtained from datasets of velocity measures characterized by a different number of observation points randomly positioned: 20, 40, 60, 80, 150, and 200.

By considering individually each analysis, some overlapping areas between the statistical scores of different ensembles are also shown in Figures 7 and 9. This means that, in some cases, a significantly different number of assimilated data can approximately lead to the same improvement. Therefore, a relatively small number of well-positioned instruments can produce an analysis almost equivalent to that produced by a network of poorly located instruments, albeit larger. In fact, some analyses obtained with only 20 ADCPs have produced results equivalent to networks with 40 or even 60 ADCPs (Figures 7 and 9).

In the second set of experiments, the positions of in situ observation instruments are identified by two elements: (1) the highest values of the projection on the velocity components of the dominant singular vectors, and (2) a fixed minimum distance among the instruments. As we mentioned in Section 2.2, the need to impose a minimum distance among instruments arises from the typical structure of the dominant singular vectors concentrated in relatively small areas.

In some experiments, not reported in this paper, we sample our DG system by extracting most measurements in the area of highly dominant SVs, but the results are even worse than those obtained with randomly positioned observations.

Distances are provided in dimensionless units, as they are divided by the barotropic Rossby deformation radius LR = (gh)1/2/f ≈ 900 km. We repeat the ensemble for testing different minimum distances from 0.04xLR to 0.28xLR.

Following the present observation strategy, the first observation point is located where the projection of the dominant SV on the velocity is maximum, and the following observation points lie farther than the chosen minimum distance.

The results of this set of experiments are shown in Figures 10 and 11, in which we compare the analyses obtained in the first assimilation window (Figure 10) and, on average, for all the assimilation windows (Figure 11).

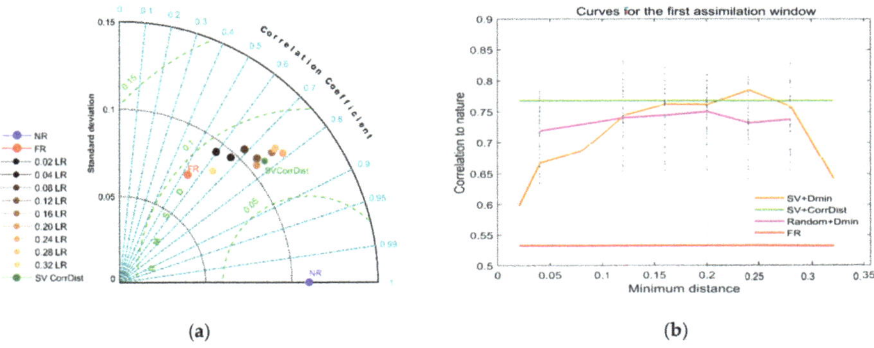

Figure 10. Representation in the Taylor diagram (**a**) and correlation graph (**b**) of the analyses obtained for the first assimilation window. Each case corresponds to a network configuration defined by the highest SVs and the minimum fixed distance, expressed as a fraction of the Rossby deformation radius. The green point and the green line correspond to the case of a variable distance based on the correlation map.

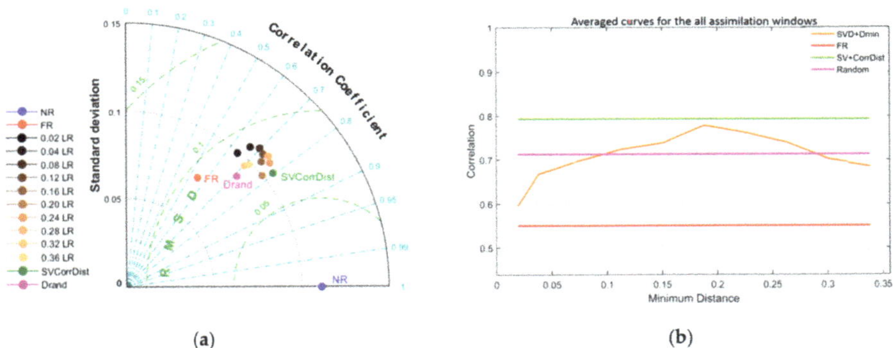

Figure 11. Taylor's diagram (**a**) and correlation graph (**b**) of the average statistics of the analyses obtained for all assimilation windows. Each case corresponds to a network configuration defined by the highest SVs and the minimum fixed distance, expressed as a fraction of the Rossby deformation radius. The green point and the green line correspond to the case of a variable distance on the correlation map.

Looking at the first assimilation window (0–5 days), the worse dataset corresponds to an imposed minimum distance of the order of the spatial resolution, which is 0.04xLR. By increasing this minimum distance, we progressively obtained better correlation up to a value approaching 0.8, in turn corresponding to a minimum separation of about 0.25xLR. The observation positions of both the worst and the best dataset are reported in Figure 12. Considering other assimilation windows, the results are quite similar: the worst correlations are usually obtained for short minimum distances, and the correlation tends to increase when we separate the observation positions. The maximum correlation is found in the range of 0.15–0.25xLR (Figure 13).

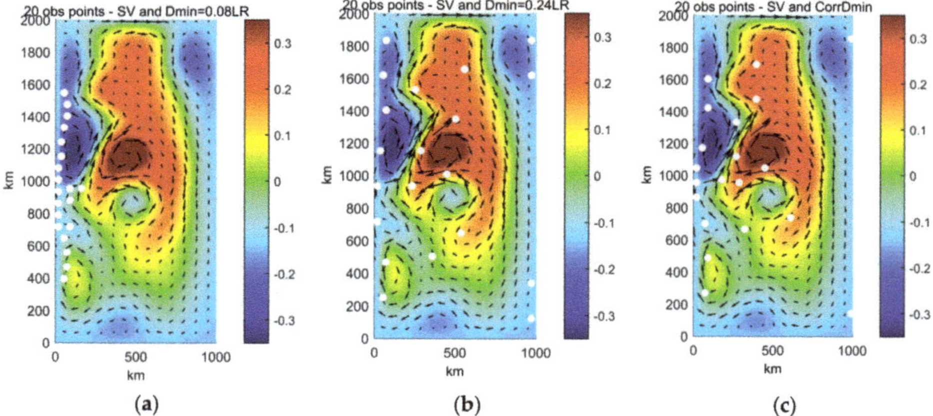

Figure 12. Observation points of dataset selected by the Singular Value Decomposition (SVD)-based procedure imposing a minimum distance of 0.08 LR (**a**), 0.24*LR (**b**), and, finally, produced by the SV-based imposing the correlation distance of Figure 10 (**c**).

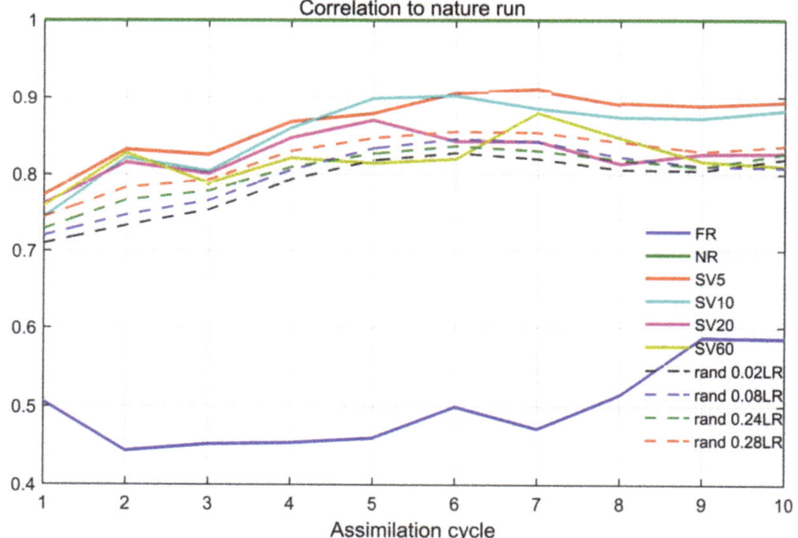

Figure 13. Correlation of analyses run to the Nature Run for the 10 cycles of the 5-day-long-assimilation windows where the observation networks were based on the correlation map of Figure 4 and the SVD computed for different optimization times: 5 days, 10 days, 20 days, and 60 days. The dashed lines represent the results for the random strategy plus a fixed minimum distance. The analyses are compared to each other and to the Free Run (blue line).

Taking the average of all curves referred to the whole set of assimilation windows, the best correlations are obtained by imposing a minimum distance of about 0.2 xLR (Figure 13).

Datasets selected by imposing a distance up to 0.08–0.1 xLR produce worse analyses than randomly selected observations; the same occurs with minimum distances larger than 0.3 xLR. Conversely, in the range between 0.1 xLR and 0.25–0.3 xLR, this selection procedure produces, on average, analyses more reliable than that corresponding to random positioning.

Finally, as a third set of experiments, we test a procedure, still based on SVD, in which we impose a maximum correlation between the time series of the observed variables at the observation positions, instead of a fixed minimum distance. Such a limitation on the correlation between the time series of model velocities at observation points is introduced by imposing a minimum distance variable over the domain, and it is computed as explained in Section 2.2.

The selection procedure for this set of experiments is based on the map in Figure 4. Starting from the observation point in which we compute the maximum value of the projection of the SV onto the velocity, we identify the minimum distance to place additional observations through such a map derived from the correlation analysis.

It is important to underline that, in this procedure, the position of observation points are uniquely determined once a correlation threshold is defined. Such a threshold should be itself a calibration parameter of the sampling strategy to be selected on the basis of the correlation value that may guarantee the best comparison between analysis and the (virtual) truth. Although the correlation threshold is not calibrated in this study, such a unique configuration of the observation points is found to be the one that gives rise, on average, to the best analysis.

The average value of such a variable correlation distance is around 300 km, which is about 0.3xLR, slightly outside the "best" minimum distance found for the second set of experiments in which a fixed separation among observations is set. However, this distance, on the map in Figure 4, varies between lower values (around 150–200 km), near the convergence area, and higher values (>600 km) near the northern, southern, and eastern edges. The highest SVs are located within the area of lower correlation distances that fall within the range of 0.15–0.20 LR that we find, empirically, as an optimal distance interval using the fixed distance criterion.

The third procedure is able, on average, to improve the quality of the analysis model with respect to both the random procedure, and to the SVD-based procedure with a fixed distance among observations.

The observation strategy is tested for repeated assimilation cycles, as in the normal procedures adopted in the operational practice. A summary representation of results is shown in Figure 13. The observation strategy based on the combination of SVD and correlation analysis (solid lines) in most cases gives the best performances with respect to any random positioning procedure (dashed lines).

We also assess the sensitivity of the results to the optimization time for the SVD computation. In this case, the difference among the datasets, in terms of correlation with respect to the NR, is not so relevant, but we find that the configuration of observation points obtained by the SVD on a shorter optimization period (5–10 days) gives the best results. In particular, the analysis corresponding to the SVD with an optimization time of five days (SV5 in Figure 13) always yields significantly better analyses than any other tested strategy.

4. Discussion and Conclusions

The marginal improvement of the reliability of an ocean forecasting system can be obtained by a proper design of the ocean-observing component.

The Singular Value Decomposition was already used by various authors in the field of Geophysical Fluid Dynamics (mainly in the atmosphere) for several purposes, including the identification of possible adaptive observation strategies. However, the analysis of the potential of this method is still rather lacking to provide effective and functional indications for the design of in situ observation networks.

In this work, we evaluate some possible SVD-based strategies to determine an optimal set of in situ observation points in the case only a limited number of observation tools are available. This situation is common in reality, given the high cost of installing and managing in situ observation networks, especially in the oceanographic field.

We compare three observation strategies aimed at reducing the forecast uncertainty obtained through an idealized Double-Gyre ocean model, with repeated analysis and forecast cycles, using the variational assimilation scheme ROMS-IS4DVar.

We first proceed to evaluate the benefit linked to the assimilation of randomly positioned observations. The assimilation algorithm in use always produces a positive improvement in the estimation of the state of the system. The effectiveness of this improvement is not straightforward, as it depends in a complex way on the number of observation points and also on the location of these points in the model domain. This is especially evident in case only a limited number of observations is available.

Having a limited number of observation tools, and looking for the combination of positions that gives maximum benefit to DA, we assume that a fundamental indication for selecting observation points can be provided by the study of the areas in which the maximum error growth occurs. SVD is an excellent method for identifying these areas. The computation of the dominant Singular Vectors (SVs), and in particular of its projection on the physical components of interest, i.e., the velocity field, can give important information about error dynamics in the limit of validity of the linear tangent model. However, as the highest values of such SVs components can be concentrated in small areas, information obtained from points too close to each other is likely to be too correlated. To avoid this effect, we test two criteria:

- A first criterion, based on a rigid distance, is able to identify an optimal separation distance, which, in this case, is equal to about one-fifth of the Barotropic Rossby Deformation Radius. Around this value we have, on average, the best skills for the analysis model compared to the Nature Run assumed as truth;
- A second criterion, based on the maximum correlation between points, adopts a variable minimum distance among observation points. This criterion defines uniquely the position of observation points and provides better results both with respect to random simulations and with respect to the former optimization criterion.

Further improvements of the last criterion can be achieved through an accurate calibration of the threshold correlation parameter. However, even when adopting a threshold parameter of the first attempt, the obtained results are, on average, better than any formerly adopted strategy.

The extension of this method to real applications must take into account other factors, such as the presence of other variables of interest or a more accurate characterization of the observation error. In cases of ocean models more complex than the ocean DG, when baroclinic effects and density variations are more important, an SVD-based observation strategy should also evaluate the projection of the dominant SV on other variables, such as temperature and salinity, as any observation strategy cannot disregard the acquisition of density profiles. The application of this method to real ocean systems will also require a careful characterization of measurement errors, estimated from the performances of real observation instruments.

Testing such criteria to the design of observation networks, as in the standard Observing System Simulation Experiments (OSSEs) used for simulating the possible benefits of observing systems, could be of great interest. Indeed, most existing ocean observation networks are not designed from the very beginning using objective criteria to optimally support analysis/forecast models. Suitable design strategies are therefore needed for both making up new observation systems and expanding the capabilities of existing observation networks in order to improve their efficiency for data assimilation.

Author Contributions: Conceptualization, M.F. and C.B.; data curation, M.F.; formal analysis, M.F.; investigation, M.F.; methodology, M.F. and C.B.; supervision, C.B.; writing—original draft, M.F.; writing—review and editing, C.B. All authors have read and agreed to the published version of the manuscript.

Funding: This research received no external funding.

Acknowledgments: The authors wish to thank their colleagues Michele Bendoni and Matteo Vannini for their support in revising the manuscript, as well as the reviewers for useful comments and suggestions.

Conflicts of Interest: The authors declare no conflict of interest.

References

1. OECD. *The Ocean. Economy in 2030*; Organisation for Economic Co-Operation and Development (OECD): Paris, France, 2016. [CrossRef]
2. She, J.; Allen, I.; Buch, E.; Crise, A.; Johannessen, J.A.; Le Traon, P.-Y.; Lips, U.; Nolan, G.; Pinardi, N.; Reissmann, J.H.; et al. Developing European operational oceanography for Blue Growth, climate change adaptation and mitigation, and ecosystem-based management. *Ocean. Sci.* **2016**, *12*, 953–976. [CrossRef]
3. Tanhua, T.; McCurdy, A.; Fischer, A.; Appeltans, W.; Bax, N.J.; Currie, K.; Deyoung, B.; Dunn, D.C.; Heslop, E.E.; Glover, L.K.; et al. What we have learned from the framework for ocean observing: Evolution of the global ocean observing system. *Front. Mar. Sci.* **2019**, *6*. [CrossRef]
4. Oke, P.R.; Larnicol, G.; Jones, E.M.; Kourafalou, V.; Sperrevik, A.K.; Carse, F.; Tanajura, C.A.S.; Mourre, B.; Tonani, M.; Brassington, G.B.; et al. Assesing the impact of observations on ocean forecasts and reanalyses: Part 2, Regional applications. *J. Oper. Oceanogr.* **2015**, *8*, s63–s79. [CrossRef]
5. Moore, A.M.; Arango, H.G.; Broquet, G.; Powell, B.S.; Weaver, A.T.; Zavala-Garay, J. The Regional Ocean Modeling System (ROMS) 4-dimensional variational data assimilation systems: Part I-System overview and formulation. *Prog. Oceanogr.* **2011**, *91*, 34–49. [CrossRef]
6. Smith, K.D.; Moore, A.M.; Arango, H. Estimates of ocean forecast error covariance derived from Hessian Singular Vectors. *Ocean. Model.* **2015**, *89*, 104–121. [CrossRef]
7. Lorenz, E.N. A study of the predictability of a 28-variable atmospheric model. *Tellus* **1965**, *17*, 321–333. [CrossRef]
8. Kantz, H.; Schreiber, T. *Nonlinear Time Series Analysis*; Cambridge University Press: Cambridge, UK, 2004.
9. Toth, Z.; Kalnay, E. *Ensemble Forecasting at NMC: The Generation of Perturbations*; NMC: Washington, DC, USA, 1993; p. 20233.
10. Ehrendorfer, M.; Tribbia, J.J. Optimal prediction of forecast error covariances through singular vectors. *J. Atmos. Sci.* **1997**, *54*, 286–313. [CrossRef]
11. Corazza, M.; Kalnay, E.; Patil, D.J.; Yang, S.-C.; Morss, R.; Cai, M.; Szunyogh, I.; Hunt, B.R.; Yorke, J.A. Use of the breeding technique to estimate the structure of the analysis "errors of the day". *Nonlinear Process. Geophys.* **2003**, *10*, 233–243. [CrossRef]
12. Palmer, T.N.; Zanna, L. Singular vectors, predictability and ensemble forecasting for weather and climate. *J. Phys. A Math. Theor.* **2013**, *46*, 254018. [CrossRef]
13. Hansen, J.A.; Smith, L.A. The role of operational contraints in selecting supplementary observations. *J. Atmos. Sci.* **2000**, *57*, 2859–2871. [CrossRef]
14. Farrell, B.F.; Ioannou, P.J. Generalized stability theory. Part I: Autonomous operators. *J. Atmos. Sci.* **1996**, *53*, 2025–2040. [CrossRef]
15. Farrell, B.F.; Ioannou, P.J. Generalized stability theory. Part II: Nonautonomous operators. *J. Atmos. Sci.* **1996**, *53*, 2041–2053. [CrossRef]
16. Wikle, C.K. Atmospheric modeling, data assimilation, and predictability. *Technometrics* **2005**, *47*, 521. [CrossRef]
17. Palmer, T.N.; Gelaro, R.; Barkmeijer, J.; Buizza, R. Singular vectors, metrics, and adaptive observations. *J. Atmos. Sci.* **1998**, *55*, 633–653. [CrossRef]
18. Buizza, R.; Montani, A. Targeting observations using singular vectors. *J. Atmos. Sci.* **1999**, *56*, 2965–2985. [CrossRef]
19. Langland, R.H.; Tóth, Z.; Gelaro, R.; Szunyogh, I.; Shapiro, M.A.; Majumdar, S.J.; Morss, R.E.; Rohaly, G.D.; Velden, C.; Bond, N.; et al. The North Pacific Experiment (NORPEX-98): Targeted observations for improved North American weather forecasts. *Bull. Am. Meteorol. Soc.* **1999**, *80*, 1363–1384. [CrossRef]
20. Buizza, R.; Cardinali, C.; Kelly, G.; Thépaut, J.-N. The value of observations. II: The value of observations located in singular-vector-based target areas. *Q. J. R. Meteorol. Soc.* **2007**, *133*, 1817–1832. [CrossRef]
21. Cardinali, C.; Buizza, R.; Kelly, G.; Shapiro, M.; Thépaut, J.-N. The value of observations. III: Influence of weather regimes on targeting. *Q. J. R. Meteorol. Soc.* **2007**, *133*, 1833–1842. [CrossRef]
22. Langland, R.H. Issues in targeted observing. *Q. J. R. Meteorol. Soc.* **2005**, *131*, 3409–3425. [CrossRef]
23. Bergot, T.; Hello, G.; Joly, A.; Malardel, S. Adaptive observations: A feasibility study. *Mon. Weather. Rev.* **1999**, *127*, 743–765. [CrossRef]

24. Gelaro, R.; Langland, R.H.; Rohaly, G.D.; Rosmond, T.E. As assessment of the singular-vector approach to targeted observing using the FASTEX dataset. *Q. J. R. Meteorol. Soc.* **1999**, *125*, 3299–3327. [CrossRef]

25. Langland, R.H.; Rohaly, G.D. *Adjoint-Based Targeting of Observations for FASTEX Cyclones*; Naval Research Lab: Monterey, CA, USA, 1996.

26. Pu, Z.X.; Kalnay, E. Targeting observations with the quasi-inverse linear and adjoint NCEP global models: Performance during FASTEX. *Q. J. R. Meteorol. Soc.* **1999**, *125*, 3329–3337. [CrossRef]

27. Bishop, C.H.; Toth, Z. Ensemble transformation and adaptive observations. *J. Atmos. Sci.* **1999**, *56*, 1748–1765. [CrossRef]

28. Szunyogh, I.; Toth, Z.; Emanuel, K.A.; Bishop, C.H.; Snyder, C.; Morss, R.E.; Woolen, J.; Marchok, T. Ensemble-based targeting experiments during fastex: The effect of dropsonde data from the lear jet. *Q. J. R. Meteorol. Soc.* **1999**, *125*, 3189–3217. [CrossRef]

29. Haidvogel, D.; Arango, H.; Budgell, W.; Cornuelle, B.; Curchitser, E.; Di Lorenzo, E.; Fennel, K.; Geyer, W.; Hermann, A.; Lanerolle, L.; et al. Ocean forecasting in terrain-following coordinates: Formulation and skill assessment of the Regional Ocean Modeling System. *J. Comput. Phys.* **2008**, *227*, 3595–3624. [CrossRef]

30. Speich, S.; Dijkstra, H.; Ghil, M. Successive bifurcations in a shallow-water model applied to the wind-driven ocean circulation. *Nonlinear Process. Geophys.* **1995**, *2*, 241–268. [CrossRef]

31. Moore, A.M.; Arango, H.G.; Di Lorenzo, E.; Cornuelle, B.; Miller, A.J.; Neilson, D.J. A comprehensive ocean prediction and analysis system based on the tangent linear and adjoint of a regional ocean model. *Ocean. Model.* **2004**, *7*, 227–258. [CrossRef]

32. Shen, J.; Medjo, T.; Wang, S. On a wind-driven, double-gyre, quasi-geostrophic ocean model: Numerical simulations and structural analysis. *J. Comput. Phys.* **1999**, *155*, 387–409. [CrossRef]

33. Kalnay, E. *Atmospheric Modeling, Data Assimilation and Predictability*; Cambridge University Press: Cambridge, UK, 2003.

34. Buehner, M.; Zadra, A. Impact of flow-dependent analysis-error covariance norms on extratropical singular vectors. *Q. J. R. Meteorol. Soc.* **2006**, *132*, 625–646. [CrossRef]

35. Golub, G.H.; Van Loan, C.F. *Matrix Computations*; Johns Hopkins University Press: Baltimore, MD, USA, 1989.

36. Taylor, K.E. Summarizing multiple aspects of model performance in a single diagram. *J. Geophys. Res. Atmos.* **2001**, *106*, 7183–7192. [CrossRef]

Publisher's Note: MDPI stays neutral with regard to jurisdictional claims in published maps and institutional affiliations.

Article

A Multivariate Balanced Initial Ensemble Generation Approach for an Atmospheric General Circulation Model

Juan Du *, Fei Zheng *, He Zhang and Jiang Zhu

International Center for Climate and Environment Science, Institute of Atmospheric Physics, Chinese Academy of Sciences, Beijing 100029, China; zhanghe@mail.iap.ac.cn (H.Z.); jzhu@mail.iap.ac.cn (J.Z.)
* Correspondence: dujuan10@mail.iap.ac.cn (J.D.); zhengfei@mail.iap.ac.cn (F.Z.)

Abstract: Based on the multivariate empirical orthogonal function (MEOF) method, a multivariate balanced initial ensemble generation method was applied to the ensemble data assimilation scheme. The initial ensembles were generated with a reasonable consideration of the physical relationships between different model variables. The spatial distribution derived from the MEOF analysis is combined with the 3-D random perturbation to generate a balanced initial perturbation field. The Local Ensemble Transform Kalman Filter (LETKF) data assimilation scheme was established for an atmospheric general circulation model. Ensemble data assimilation experiments using different initial ensemble generation methods, spatially random and MEOF-based balanced, are performed using realistic atmospheric observations. It is shown that the ensembles integrated from the balanced initial ensembles maintain a much more reasonable spread and a more reliable horizontal correlation compared with the historical model results than those from the randomly perturbed initial ensembles. The model predictions were also improved by adopting the MEOF-based balanced initial ensembles.

Keywords: MEOF; initial ensemble; ensemble spread; LETKF; data assimilation

Citation: Du, J.; Zheng, F.; Zhang, H.; Zhu, J. A Multivariate Balanced Initial Ensemble Generation Approach for an Atmospheric General Circulation Model. *Water* **2021**, *13*, 122. https://doi.org/10.3390/w13020122

Received: 30 October 2020
Accepted: 29 December 2020
Published: 7 January 2021

Publisher's Note: MDPI stays neutral with regard to jurisdictional claims in published maps and institutional affiliations.

1. Introduction

The ensemble Kalman filter (EnKF) data assimilation approach was introduced by Evensen in 1994 [1], which is a Monte-Carlo approach and has the potential for efficient use on parallel computers with large-scale geophysical models [2–8]. The EnKF method uses an ensemble of model forecasts to estimate the background error covariances and optimizes the background with the available observations. So it is easy to implement (no adjoint models are required compared with the three-dimension variational data assimilation [2,9]) and handles strong non-linearities better than other known Kalman filter techniques for large-scale problems [10].

EnKF was first applied to an atmospheric model by Houtekamer and Mitchell [11]. After that, it has rapidly become a promising choice for the operational numerical weather prediction systems. The square root filter (SRF) method of EnKF without perturbed observations (deterministic filters) was proposed by assimilating the observations serially [12,13], and then the EnKF method with perturbed observations (stochastic filters) was applied to a pre-operational system [14]. A local ensemble Kalman filter (LEKF) method that assimilates observations simultaneously was proposed by Ott et al. [15]. Furthermore, the local ensemble transform Kalman filter (LETKF) which uses the ensemble transform Kalman filter (ETKF) approach was proposed to further accelerate LEKF [16,17]. The LETKF assimilates observations within a spatially physical local volume at each model grid point simultaneously and does not require an orthogonal basis which significantly enhances the computational efficiency with parallel implementation [17,18].

For the initial perturbation generation, several kinds of methods have been developed, such as error breeding [19], singular vectors [20], perturbed observations [21] and random perturbations [5,21–23]. The performances of these methods were illustrated in several numerical weather prediction models with different complexities [24–27]. Zheng and

89

Zhu proposed a multivariable empirical orthogonal function (MEOF) based model error perturbation to generate perturbed model errors and then applied it to a global spectral atmospheric model with real observations [28,29]. It should be realized that how to maintain the physical relationships of the different model variables induced by the initial perturbations and how to provide a reasonable background covariance are still an important problem for the ensemble data assimilation process.

In this work, the local ensemble transform Kalman filter approach has been implemented for an atmospheric general circulation model developed by the Institute of Atmospheric Physics (IAP AGCM version 4). A MEOF based balanced perturbation generation method is adopted for generating the initial ensembles, compared with the spatially random perturbation method [5]. The remainder of this paper is structured as follows: In Sections 2 and 3, the forecast model and the LETKF data assimilation scheme are briefly described respectively. In Section 4, the implementation of the initial perturbation generation scheme based on the multivariate empirical orthogonal function (MEOF) is introduced. In Section 5, the spatially-correlated random perturbation scheme and the MEOF-based balanced perturbation scheme are both applied to the AGCM model results to generate the initial ensemble. The ensemble spread and horizontal correlation of the initial ensembles are compared for the two methods. And the LETKF data assimilation scheme is applied to the AGCM model with the conventional observation data. The characteristics and effects of the random and MEOF based initial ensemble generation methods are illustrated respectively. The data assimilation results using the two different initial ensembles are also shown in this section. Summary and conclusions are drawn in the final section.

2. The Forecast Model

The atmospheric general circulation model used here is the IAP AGCM version 4 as a component of the Chinese Academy of Sciences (CAS) earth system model (ESM). The model was applied to the simulation of atmospheric circulations and climate, such as summer precipitation and monsoons [30,31]. It is a global grid-point model using finite-difference scheme with a terrain-following σ coordinate. A latitude-longitude grid with Arakawa's C grid staggering is used in the horizontal discretization [32–34]. The formulation of the governing equations and the finite-difference schemes have several novel features in the IAP AGCM. The model equations are based on the baroclinic primitive equations with subtraction of standard stratification. The purpose of subtracting the standard stratification in the dynamical core is to reduce truncation errors, especially over regions of high terrain. And the IAP model conserves total available energy (sum of kinetic energy, the available potential energy, and the available surface potential energy) rather than total energy. To maintain the conservation of the total available energy, a variable substitution method named the IAP transform is adopted in the numerical design. The model resolution adopted here is 1 degree by 1 degree. The nonlinear iterative time integration method described in [35] is used in the model. The timestep adopted in the numerical simulation here is 1200 seconds. The prognostic model variables are temperature, surface pressure, wind velocity and specific humidity.

3. Data Assimilation Scheme

The Local Ensemble Transform Kalman Filter (LETKF) algorithms used here are based on the work of Hunt et al. [17]. An important advantage of LETKF schemes compared to EnKF is their efficiency in parallel computing. Because LETKF separates the entire global grid into independent local regions, ideally they have the total parallel efficiency [18]. In this section, we introduce the main idea of LETKF briefly.

The ensemble members are defined as $x_i \in \Re^n (i = 1, \cdots, N)$, where N is the ensemble size and n is the dimension of the model state. The ensemble matrix X can be constructed by the model states of the ensemble as:

$$X = (x_1, x_2, \cdots, x_N) \in \Re^{n \times N} \tag{1}$$

The anomaly matrix is

$$X' = X - \overline{X} \tag{2}$$

where $\overline{X}$ is the ensemble mean vector.

To update analysis states at every grid point, the LETKF assimilates only observations within a certain distance from each grid point. Here we use the subscript l to denote a quantity defined on such a local region centered at an analysis grid point. The analysis mean is

$$\overline{x}_l^a = \overline{x}_l + x_l P_l^a (Y_l)^T R_l^{-1} (y_l^o - y_l) \tag{3}$$

And the analysis error covariance matrix P_l^a is

$$P_l^a = [(Y_l)^T R_l^{-1} Y_l + (N-1)\mathbf{I}/\rho]^{-1} \tag{4}$$

where R_l is the observation error covariance matrix. The observation vector is $y \in \Re^m$. $Y = (y_1, \cdots, y_N)$ and H is the observation operator which interpolates the model state to the observation space. ρ is the multiplicative inflation factor. Within a local region, space localization is carried out by multiplying the inverse observation error covariance matrix with a factor that decays from one to zero as the distance of the observations from the analysis grid point increases [36].

4. Multivariable Balanced Initial Perturbation Scheme

Based on the multivariate empirical orthogonal function MEOF [28], a MEOF-based multivariable initial perturbation method is adopted here to generate a balanced initial ensemble state for the LETKF data assimilation, which can make the ensemble members maintain a reasonable spread as the forecast model integrates. For the MEOF analysis, the snapshots of all the model variables are put in one single vector to make the EOF analysis, instead of making EOF analysis for the model variables individually. The spatial distribution of the model snapshots, which are derived from the MEOF analysis, and the 3-D random perturbation are combined together to generate a balanced perturbation field. The detailed implementation steps of the method are described as follows:

$$Q_i(x,y,z,v) = D(x,y,z,v) + \sum_{j=1}^{N_m} \sigma_j(z,v)\phi_j(x,y,z,v)\omega_{i,j}, i = 1,\ldots,N \tag{5}$$

where $Q_i(x,y,z,v)$ represents the generated initial perturbation field for the ith ensemble member, and $D(x,y,z,v)$ represents the initial model state. $\sigma_j(z,v)$ represents the standard deviation of model variables in different model layers, which can be calculated from the time coefficients of the MEOF analysis. N_m is the chosen mode number according to the MEOF analysis. $\phi_j(x,y,z,v)$ is the analyzed spatial MEOF mode of the model state variables in different layers. $\omega_{i,j}$ is a one-dimension random vector with a mean equal to 0 and variance equal to 1, and the random vectors $\omega_{i,j}(j=1,\ldots,N_m)$ should be independent to make the MEOF modes orthogonal. x, y and z represent the 3-D coordinate, v represents different model variables, and N is the ensemble size.

In practice, we could derive the departures of the model integration results from their average in each model layer first to generate the balanced initial perturbation fields. The standard deviations $\sigma(z,v)$ of the model variables in each model layer could be calculated to normalize the model variables in all the model layers. The MEOF analysis is performed for the normalized model variables and the spatial modes $\phi(x,y,z,v)$ could be obtained. Finally, we can apply the above equation to generate the initial ensemble perturbation fields. Because the perturbations are a combination of the spatial distribution of all the model variables, the initial ensembles were generated with a reasonable consideration of the physical relationships between different model variables. Then, we can add the derived MEOF based perturbations to the initial state of the model, which are the model's prognostic variables (i.e., ps, U, V, T, q). After the initial ensembles are generated, we integrate the model for six hours and use the six-hour model forecast as the analysis samples, because it is crucial to check whether the ensemble spread and the spatial correlation at the first analysis time maintain reasonable.

5. Data Assimilation Experiments

Two different initial ensemble generation methods are tested for the LETKF data assimilation of the AGCM. One method is the spatially-correlated random perturbation scheme [5], and the other one is the MEOF-based balanced perturbation scheme. For the two initial perturbation schemes, 80 ensemble members are adopted for the ensemble data assimilation process. The observational data adopted here are the global upper air and surface weather observation data in PREPBUFR format, which are usually used as the conventional observation data for the data assimilation system. The data include land surface, marine surface, radiosonde, pibal and aircraft reports, profiler and radar derived winds, satellite wind data and so on. The data can include pressure, geopotential height, temperature, dew point temperature, wind direction and speed. The conventional observations are grouped into a time window of 6 hours, which are centered on the analysis time, and then are assimilated into the model every 6 hours from 1 January to 10 January 2004, which are at 0000, 0600, 1200 and 1800 UTC. An example figure of the conventional observation data of the surface temperature is shown in Figure 1.

As we can find out in the model integration process, the integration of the temperature variable over time will also influence the integration of the other model variables. So for the generation of the randomly perturbed initial ensemble, we just add a 3-D random noise of a certain magnitude (1% of the magnitude of T) to the temperature variable of the atmosphere general circulation model at all layers, following Evensen's idea [5]. The random perturbation is generated with a horizontal correlation scale of 2000 km and a vertical correlation scale of 1000 km, as well as a relativity of 0.8 between two adjcint layers. For the generation of the MEOF-based perturbed initial ensemble, we implement the multivariable balanced initial perturbation scheme as described in Section 4. The spatial distribution of the model snapshots derived from the MEOF analysis and the 3-D random perturbation are combined to generate a balanced perturbation field.

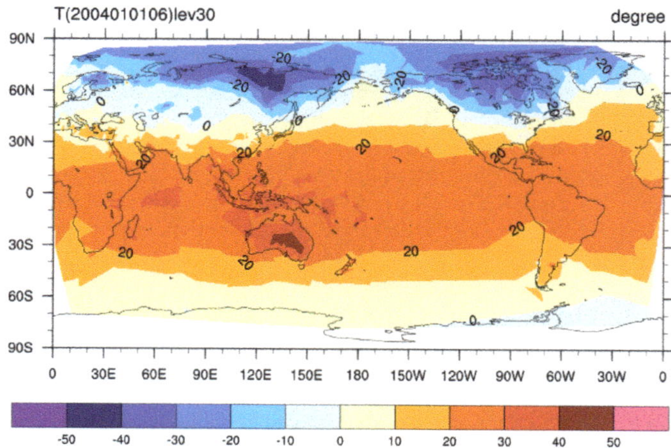

Figure 1. Conventional observation data of temperature at the surface layer at 06UTC 20040101.

5.1. The MEOF Analysis Results

For the MEOF analysis and the MEOF-based perturbation generation, the AGCM is integrated from 1 January to 31 March 2004 to generate the six-hour model forecast outputs. A total of 360 snapshots are adopted to make the MEOF analysis. Compared to the EOF function analysis for each individual model variable, the MEOF function analysis combines all the model variables in one vector. Figure 2 shows the variance contributions of the first 24 modes for the MEOF analysis of the surface pressure. The total variance contribution of the first 16 MEOF modes have been more than 99%. So the first 20 MEOF modes are adopted to generated the balanced perturbation fields. The spatial distribution and the time coefficients of the first three MEOF modes of the surface pressure (Ps) is shown in Figure 3. Similarly, we can see the detailed MEOF analysis results of the temperature (T) at the surface layer in Figures 4 and 5. The total variance contribution of the first twenty MEOF modes have been also more than 99%.

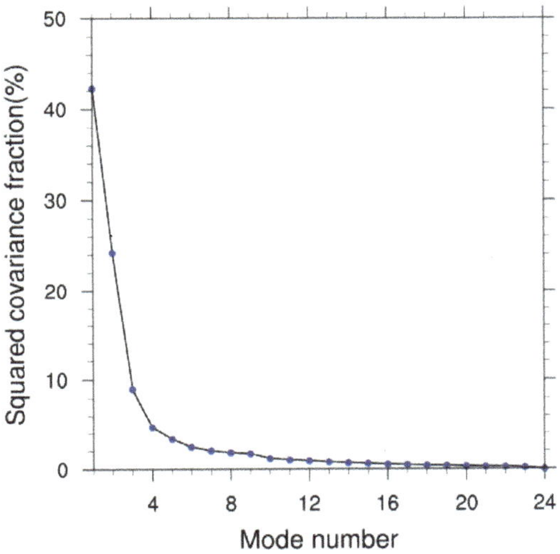

Figure 2. Variance contributions of the first 24 modes from the MEOF analysis of Ps.

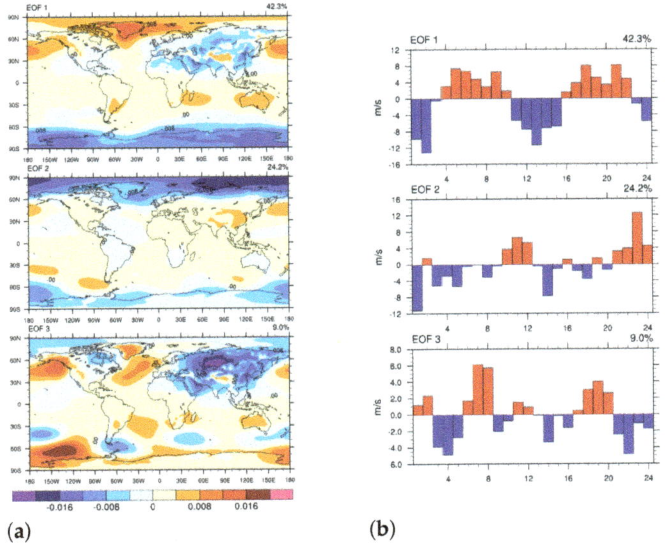

(**a**) (**b**)

Figure 3. The spatial distribution (**a**) and the time coefficients (**b**) of the first three modes of the MEOF analysis of Ps.

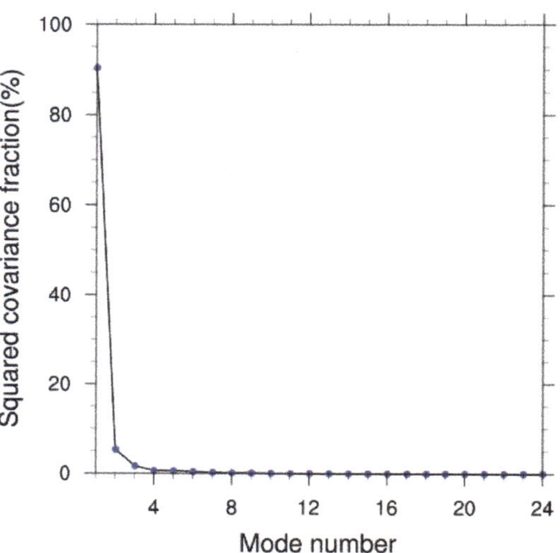

Figure 4. Variance contributions of the first 24 modes from the MEOF analysis of T.

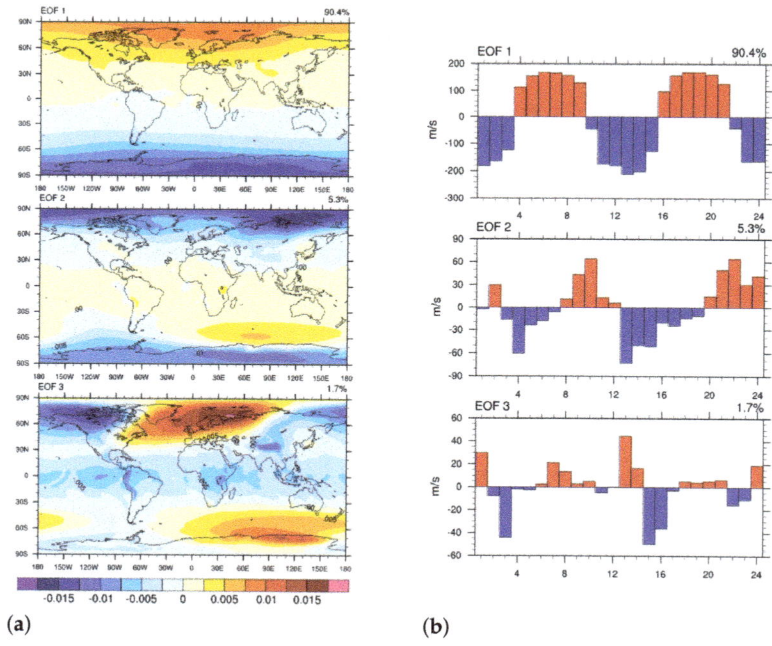

(**a**)   (**b**)

Figure 5. The spatial distribution (**a**) and the time coefficients (**b**) of the first three modes of the MEOF analysis of T.

5.2. Ensemble Spread

To verify the quality of the generated initial ensemble, it's essential to compare the ensemble spread of the initial ensemble and the model outputs after six-hour integration, which is at 06UTC 1 January 2004. A reasonable ensemble spread should represent well the distribution of the forecast uncertainties before the assimilation took place, and a larger ensemble spread can result in a Kalman gain that reasonably draws the analysis closer to the observations [28]. For the random perturbation and the MEOF balanced perturbation scheme, the ensemble spreads of T both decrease after six-hour integration compared with the initial ensemble spread, as shown in Figures 6 and 7. The difference is that the ensemble spread of the MEOF balanced perturbed ensemble decreases much less than that of the randomly perturbed ensemble. The averaged spread of the randomly perturbed initial ensemble of the temperature at the surface layer is about 3.3 degree, which decreases to about 1.4 degree after six-hour integration. As a contrast, the averaged ensemble spread of the MEOF-based balanced initial perturbation of the temperature at the surface layer is about 7.2 degree, which decreases to about 6.1 degree after six-hour integration. It's shown that the MEOF balanced perturbation could maintain the ensemble spread more reasonable, which is very important for the data assimilation process.

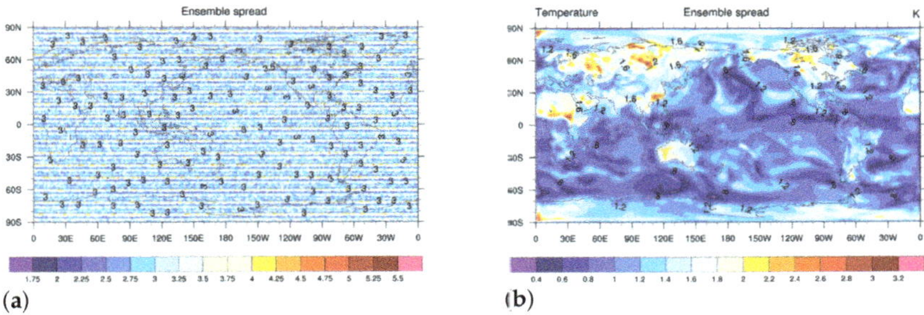

Figure 6. The initial (**a**) and the 6-h integration (**b**) ensemble spread of the randomly perturbed T.

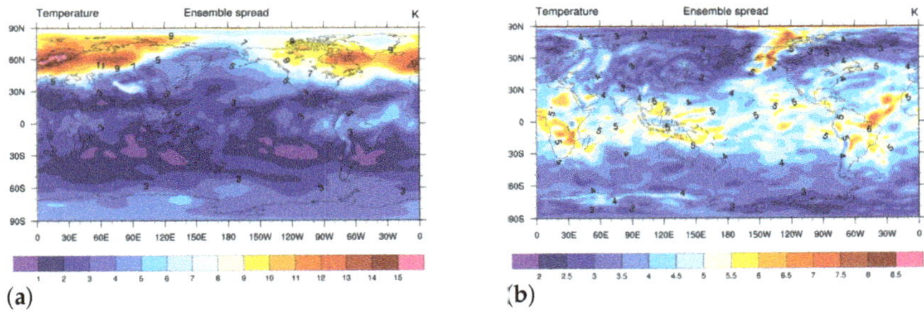

Figure 7. The initial (**a**) and the 6-h integration (**b**) ensemble spread of the MEOF perturbed T.

5.3. Horizontal Correlation

Take the surface pressure as example, we calculated the horizontal correlation of four locations for both the randomly perturbed initial ensemble and the MEOF-based balanced initial ensemble. The four locations are chosen as (67.5 E, 33.31 N), (90 E, 68.74 S), (178.59 E, 0.71 S) and (61.87 W, 55.98 N). Figure 8 shows the historical horizontal correlations of the surface pressure at the chosen four locations. The historical results include the six-hour model integration outputs from 1 August to 31 October 2004. Figure 9 shows the horizontal correlations of the randomly perturbed ensemble of the surface pressure at the chosen four locations. The ensembles used to calculate the horizontal correlation is the six-hour forecast of the MEOF-based perturbed initial ensemble. Figure 10 shows the horizontal correlations of the MEOF-based perturbed ensemble of the surface pressure at the chosen four locations. The ensembles used to calculate the horizontal correlation is the six-hour forecast of the randomly perturbed initial ensemble. We can see that the horizontal correlations of the MEOF-based perturbed ensemble of the surface pressure are much more similar to the historical horizontal correlations of the model integration, compared with the horizontal correlations of the randomly perturbed ensemble. The horizontal correlations of the randomly perturbed ensemble have the normal oval shape and can't represent the historical characteristics in the middle and high latitude area. It's shown that the ensembles generated from the MEOF perturbations could represent the historical horizontal correlations better. Similar conclusions could be driven for the other state variables, such as the temperature, the wind velocity and humidity.

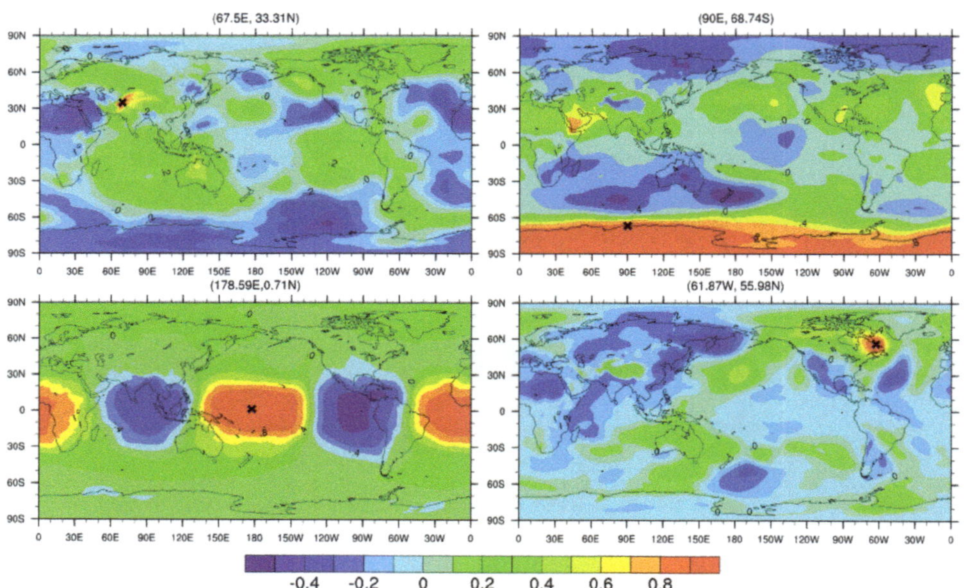

Figure 8. The historical horizontal correlation of the model integration results at four locations.

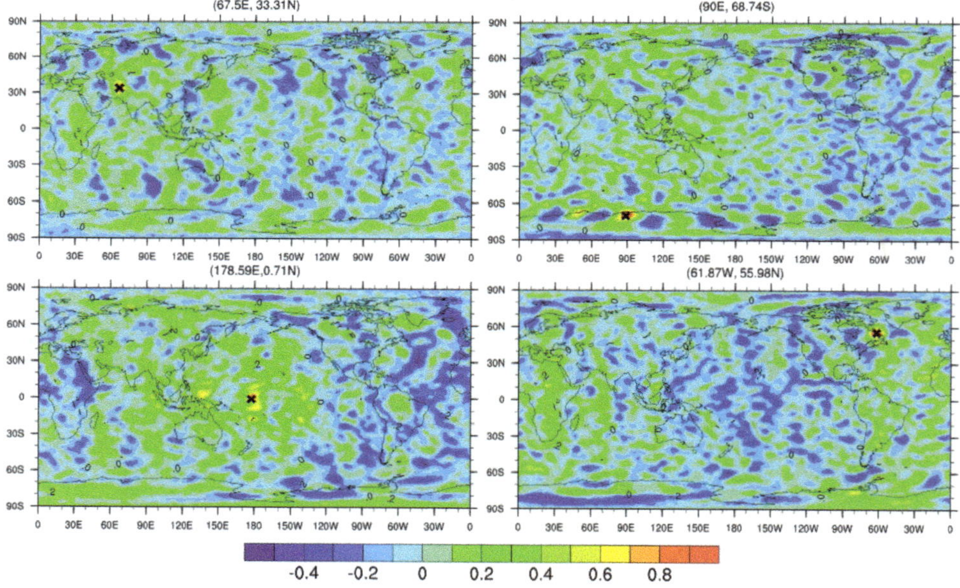

Figure 9. The horizontal correlation of the randomly perturbed ensemble at four locations.

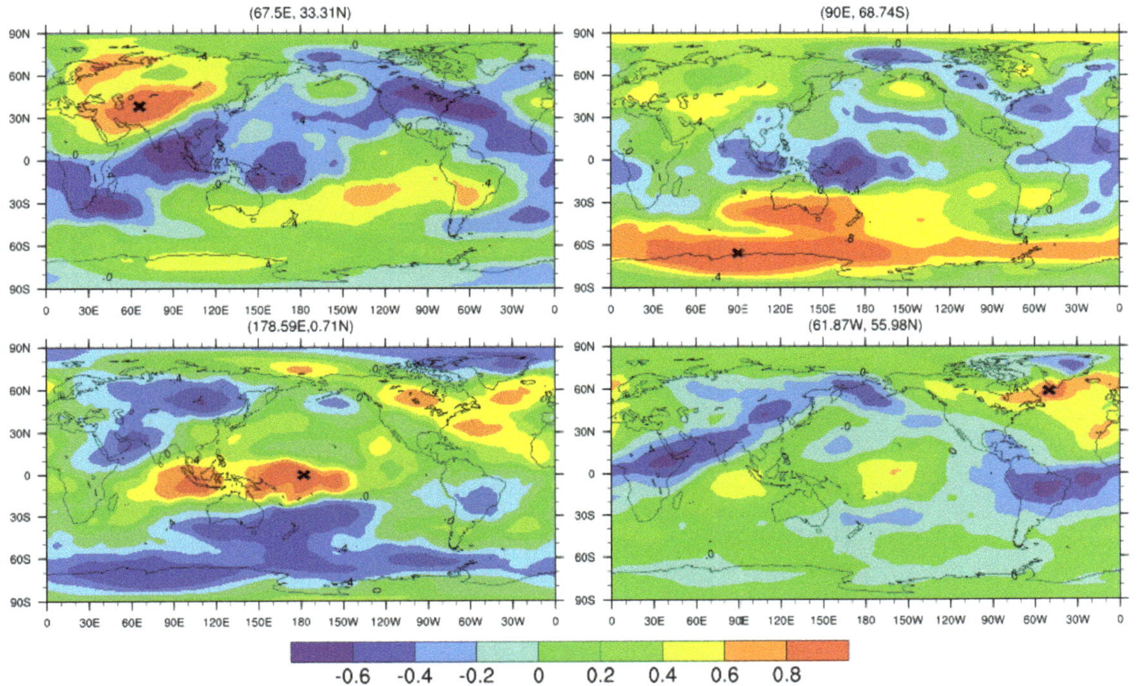

Figure 10. The horizontal correlation of the MEOF based perturbed ensemble at four locations.

5.4. LETKF Data Assimilation Results

The LETKF data assimilation scheme is applied to the atmospheric general circulation model using 80 ensembles. For the initial ensemble generation, the spatially-correlated random perturbation scheme and the MEOF-based balanced perturbation scheme are implemented and compared from several aspects, such as the ensemble spread and the horizontal correlation. We can see that the initial ensemble generated from the MEOF-based balanced perturbation has a better performance, as the ensemble forecasted from the MEOF-based perturbed initial ensemble could maintain a better spread and their horizontal correlation is more compatible with the horizontal correlation of the historical model output. Here, we adopted the MEOF-based perturbed initial ensemble to start the data assimilation process. The observation adopted here is the six-hour conventional observation data starting from 06UTC 1 January 2004. The observation data of the temperature, the meridional wind and the zonal wind have been assimilated into the AGCM. Figure 11 shows the root mean square error(RMSE) of the LETKF data assimilation results of the surface temperature for the first six data assimilation times, compared with the conventional observation data (see Figure 1). It seems that the RMSE of the data assimilation results derived from both the randomly and MEOF-based perturbed initial ensemble is smaller than the RMSE of the control model, which means the initial ensembles generated from both the two methods worked during the data assimilation process. It's also shown that the RMSE of the data assimilation results derived from the MEOF-based perturbed initial ensemble is smaller than those derived from the randomly perturbed initial ensemble. Because the initial ensemble generated by the MEOF-based perturbation has better physical relationships between the model variables, the data assimilation effect is further improved. The LETKF data assimilation also improved the meridional and zonal wind result compared to the observation (figures not shown).

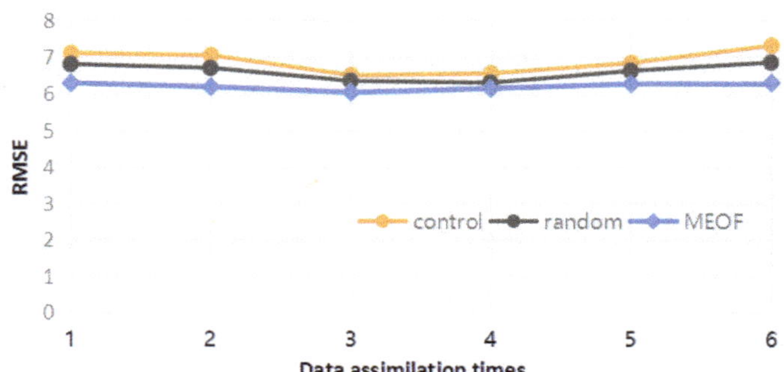

Figure 11. The RMSE of the LETKF data assimilation results compared with the observation data for the surface temperature.

6. Conclusions

Based on the multivariate empirical orthogonal function (MEOF) method, a multivariate balanced initial ensemble generation method was applied to the ensemble data assimilation scheme. The initial ensembles were generated with a reasonable consideration of the physical relationships between different model variables. For the initial ensemble generation, the spatially-correlated random perturbation scheme and the MEOF-based balanced perturbation scheme are implemented and compared from several aspects, such as the ensemble spread and the horizontal correlation. From the analysis of ensemble spread and the horizontal correlation, we can see that the initial perturbations generated based on the MEOF method are much more effective considering they will not decay rapidly as the model integrates. The ensembles integrated from the initial ensemble generated from the MEOF-based perturbations will maintain a much more reasonable spread and a more reliable horizontal correlation than those from the randomly perturbed initial fields. The Local Ensemble Transform Kalman Filter (LETKF) data assimilation scheme was established for an atmospheric general circulation model. Ensemble data assimilation experiments using different initial ensemble generation methods, spatially random and MEOF-based balanced, are performed using realistic atmospheric observations. The model predictions were also improved by adopting the MEOF-based balanced multivariate initial ensembles. At the present, only the conventional observation data is assimilated into the AGCM. More data assimilation experiments with the LETKF scheme using the satellite observation data will be made in the future research.

Author Contributions: Conceptualization, J.D.; methodology, J.D. and F.Z.; software, J.D. and H.Z.; validation, J.D.; formal analysis, J.D.; writing—original draft preparation, J.D.; writing—review and editing, J.D., F.Z. and J.Z.; visualization, J.D.; supervision, J.Z.; funding acquisition, J.D., F.Z., H.Z. and J.Z. All authors have read and agreed to the published version of the manuscript.

Funding: This work was supported by the National Key Research and Development Program of China (Grant No. 2018YFA0605703), the National Natural Science Foundation of China (Grant No. 41776041; 41630530), the Key Research Program of Frontier Sciences, CAS (Grant No. ZDBS-LY-DQC010), the National Natural Science Foundation of China (Grant No. 41876012; 41861144015) and the Strategic Priority Research Program of the Chinese Academy of Sciences (Grant No. XDB42000000).

Conflicts of Interest: The authors declare no conflict of interest.

References

1. Evensen, G. Sequential data assimilation with a nonlinear quasi-geostrophic model using Monte Carlo methods to forecast error statistics. *J. Geophys. Res.* **1994**, *99*, 10143–10162. [CrossRef]
2. Blum, J.; Le Dimet, F.X.; Navon, I.M. *Data Assimilation for Geophysical Fluids*; Chapter in Computational Methods for the Atmosphere and the Oceans, Volume 14, Special Volume of Handbook of Numerical Analysis; Temam, R., Tribbia, J., Eds.; Elsevier Science Ltd.: New York, NY, USA, 2009; 784p, ISBN 978-0-444-51893-4.

3. Burgers, G.; Evensen, G.; Leeuwen, P.J. On the Analysis Scheme in the Ensemble Kalman Filter. *Mon. Weather Rev.* **1998**, *126*, 1719–1724. [CrossRef]
4. Evensen, G.; Leeuwen, P.J. Assimilation of Geosat Altimeter Data for the Agulhas Current using the Ensemble Kalman Filter with a Quasi-Geostrophic Model. *Mon. Weather Rev.* **1996**, *124*, 85–96. [CrossRef]
5. Evensen, G. The ensemble Kalman filter: Theoretical formulation and practical implementation. *Ocean Dyn.* **2003**, *53*, 343–367. [CrossRef]
6. Kalnay, E. *Atmospheric Modeling, Data Assimilation and Predictability*; Cambridge University Press: London, UK, 2003; 341p.
7. Nerger, L.; Hiller, W.; Schroter, J. PDAF—The Parallel Data Assimilation Framework: Experiences with Kalman filtering. In *Use of High Performance Computing in Meteorology-Proceedings of the 11 ECMWF Workshop*; Zwieflhofer, W., Mozdzynski, G., Eds.; World Scientific: Reading, UK, 2005; pp. 63–83.
8. Du, J.; Zhu, J.; Fang, F.; Pain, C.C.; Navon, I.M. Ensemble data assimilation applied to an adaptive mesh ocean model. *Int. J. Numer. Meth. Fluids* **2016**, *82*, 997–1009. [CrossRef]
9. Du, J.; Navon, I.M.; Zhu, J.; Fang, F.; Alekseev, A.K. Reduced order modeling based on POD of a parabolized Navier-Stokes equations model II: Trust region POD 4D VAR data assimilation. *Comput. Math. Appl.* **2013**, *65*, 380–394. [CrossRef]
10. Sakov, P.; Oliver, D.S.; Bertino, L. An Iterative EnKF for Strongly Nonlinear Systems. *Mon. Weather. Rev.* **2011**, *140*, 1988–2004. [CrossRef]
11. Houtekamer, P.L.; Mitchell, H.L. Data assimilation using an ensemble Kalman filter technique. *Mon. Wea. Rev.* **1998**, *126*, 796–811. [CrossRef]
12. Tippett, M.K.; Anderson, J.L.; Bishop, C.H.; Hamill, T.M.; Whitaker, J.S. Ensemble square root filters. *Mon. Wea. Rev.* **2003**, *131*, 1485–1490. [CrossRef]
13. Whitaker, J.S.; Hamill, T.M. Ensemble data assimilation without perturbed observations. *Mon. Wea. Rev.* **2002**, *130*, 1913–1924. [CrossRef]
14. Houtekamer, P.L.; Mitchell, H.L.; Pellerin, G.; Buehner, M.; Charron, M.; Spacek, L.; Hansen, B. Atmospheric data assimilation with an ensemble Kalman filter: Results with real observations. *Mon. Wea. Rev.* **2005**, *133*, 604–620. [CrossRef]
15. Ott, E.; Hunt, B.R.; Szunyogh, I.; Zimin, A.V.; Kostelich, E.J.; Corazza, M.; Kalnay, E.; Patil, D.J.; Yorke, J.A. A local ensemble Kalman filter for atmospheric data assimilation. *Tellus* **2004**, *56*, 415–428. [CrossRef]
16. Bishop, C.H.; Etherton, B.; Majumdar, S.J. Adaptive sampling with the ensemble transform Kalman filter. Part I: Theoretical aspects. *Mon. Wea. Rev.* **2001**, *129*, 420–436. [CrossRef]
17. Hunt, B.; Kostelich, E.; Syzunogh, I. Efficient data assimilation for spatiotemporal chaos: A local ensemble transform Kalman filter. *Physica D* **2007**, *230*, 112–126. [CrossRef]
18. Miyoshi, T.; Yamane, S. Local ensemble transform Kalman filtering with an agcm at a T159/L48 resolution. *Mon. Wea. Rev.* **2007**, *135*, 3841–3861. [CrossRef]
19. Toth, Z.; Kalnay, E. Ensemble forecasting at NCEP and the breeding method. *Mon. Wea. Rev.* **1997**, *125*, 3297–3319. [CrossRef]
20. Buizza, R.; Palmer, T.N. The singular-vector structure of the atmospheric global circulation. *J. Atmos. Sci.* **1995**, *52*, 1434–1456. [CrossRef]
21. Houtekamer, P.L.; Derome, J. Methods for ensemble prediction. *Mon. Wea. Rev.* **1995**, *123*, 2181–2196. [CrossRef]
22. Du, J.; Mullen, S.L.; Sanders, F. Short-range ensemble forecasting of quantitative precipitation. *Mon. Wea. Rev.* **1997**, *125*, 2427–2459. [CrossRef]
23. Wan, L.; Zhu, J.; Bertino, L.; Wang, H. Initial ensemble generation and validation for ocean data assimilation using HYCOM in the Pacific. *Ocean Dyn.* **2008**, *58*, 81–99. [CrossRef]
24. Hamill, T.M.; Snyder, C.; Morss, R.E. A comparison of probabilistic forecasts from bred, singular-vector, and perturbed observation ensembles. *Mon. Wea. Rev.* **2000**, *128*, 1835–1851. [CrossRef]
25. Magnusson, L.; Nycander, J.; Kallen, E. Flow-dependent versus flow-independent initial perturbations for ensemble prediction. *Tellus* **2009**, *61A*, 194–209. [CrossRef]
26. Wang, X.; Bishop, C.H. A comparison of breeding and ensemble transform Kalman filter ensemble forecast schemes. *J. Atmos. Sci.* **2003**, *60*, 1140–1158. [CrossRef]
27. Zupanski, M.; Fletcher, S.J.; Navon, I.M.; Uzunoglu, B.; Daescu, D. Initiation of ensemble data assimilation. *Tellus* **2006**, *58A*, 159–170. [CrossRef]
28. Zheng, F.; Zhu, J. Balanced multivariate model errors of an intermediate coupled model for ensemble Kalman filter data assimilation. *J. Geophys. Res.* **2008**, *113*, C07002. [CrossRef]
29. Zheng, F.; Zhu, J. A multivariate empirical orthogonal function-based scheme for the balanced initial ensemble generation of an ensemble kalman filter. *Atmos. Ocean. Sci. Lett.* **2010**, *3*, 165–169.
30. Lin, Z.; Zeng, Q. Simulation of East Asian summer monsoon by using an improved AGCM. *Adv. Atmos. Sci.* **1997**, *14*, 513–526.
31. Zeng, Q.; Yuan, C.; Li, X.; Zhang, R.; Yang, F.; Zhang, B.; Lu, P.; Bi, X.; Wang, H. Seasonal and extraseasonal predictions of summer monsoon precipitation by GCMs. *Adv. Atmos. Sci.* **1997**, *14*, 163–176.
32. Zeng, Q.; Zhang, X.H.; Liang, X.Z.; Yuan, C.G.; Chen, S.F. *Documentation of IAP Two-Level Atmospheric General Circulation Model*; Rep. DOE/ER/60314-H1 TR044; Department of Energy Tech., State Univ. of New York: Stony Brook, NY, USA, 1989; 383p.
33. Zhang, H.; Zhang, M.; Zeng, Q. Sensitivity of simulated climate to two atmospheric models: Interpretation of differences between dry models and moist models. *Mon. Weather. Rev.* **2013**, *141*, 1558–1576. [CrossRef]

34. Zhang, H. Development of IAP Atmospheric General Circulation Model Version 4.0 and Its Climate Simulations (in Chinese). Ph.D. Thesis, Institute of Atmospheric Physics, Chinese Academy of Sciences, Beijing, China, 2009.
35. Zuo, R.; Zhang, M.; Zhang, D.; Wang, A.; Zeng, Q. Designing and climatic numerical modeling of 21-level AGCM (IAP AGCM-III). Part I: Dynamical framework (in Chinese). *Chin. Atmos. Sci.* **2004**, *28*, 659–674.
36. Shin, S.; Kang, J.S.; Jo, Y. The Local Ensemble Transform Kalman Filter (LETKF) with a Global NWP Model on the Cubed Sphere. *Pure Appl. Geophys.* **2016**, *173*, 2555–2570. [CrossRef]

MDPI

St. Alban-Anlage 66

4052 Basel

Switzerland

Tel. +41 61 683 77 34

Fax +41 61 302 89 18

www.mdpi.com

Water Editorial Office

E-mail: water@mdpi.com

www.mdpi.com/journal/water